The Revelation

The *"Peace Machine"* Revealed &

The Importance of Market-Fences &

The UNIFIED FIELD THEORY Explained

Or

God is Love

By Karl Spain

&

Dr. Mark Richards
Consultant, Interpreter and Confidant

Published by Peace Machine Press

www.TheRevelationBook.com

Copyright © 2009 by Karl Spain
All rights reserved. No part of this book may be reproduced, scanned, or distributed in any printed or electronic form without permission.

ISBN 978-0-578-02064-8

Printed in the United States of America

This book is a not a work of fiction. Any names, characters, incidents and/or facts are all intended to be accurate, and any resemblance to actual persons, countries, and religions is purposeful. Nevertheless -- if the Author gives any offense to any such person or institution, such offense is unintended, coincidental, and regretted.

First Edition
Second Printing

God is love

Dedications

This book is dedicated with Love
to my two children -- Max and Merrill.

My beloved sister Amber, my love Kerry,
My family, especially my west coast cuz, Maura,
As well as my angels: Dorothy, Aunt Gladys & Ed B.

And all my loyal, loving friends, but in particular:
Jon & David, Roz, Sharon, Trinka, The Tallman, Aris,
Barbara & Ralph, Melinda, Leo, Kevin, Jeff, Scott & Laura,
Ishan, Eddie, and even Evan, although he never calls back.

To my editors & copy editors:
Dave Sampselle (a man I am also fortunate to call a friend)
& Jane Touzalin, do I ever need (and love) you both!

To all my friends at Briggs Welding, Panera Bread, Brothers Pizza, and California Tortilla in Charles Town, West Virginia, where *the bulk of this book was written*, thank you for the food & friendship.

To those to whom I owe a mention and have omitted one,
my <u>sincerest</u> apologies.

To every reader:
Without our angels and love,
We all stand alone *with* evil.

***This book is written in the hope
of advancing human knowledge.***

> A special thank you to Dr. Mark
> Richards, a man I love, admire, and will
> never be able to thank enough for the
> wonderful work he did on
> *The Revelation.*
> To know and respect another man's
> mind is a very special
> bond, one I am honored to
> have shared with him.

In that vein, I want to add a special thank-you to
Thorbjørn Schibelfeldt, The Dane,
who actually had nothing to do with writing
this book, but had a lot to do with my ability
to write it.

Please do not skip the Prelude or Introduction.

The Revelation -- Table of Contents

Prelude	7
Introduction	12

Book I -- The Peace Machine Hypothesis

Chapter 1	Let Down Your Guard	24
Chapter 2	Super Darwinism	27
Chapter 3	A New Brain Inside the Old One	47
Chapter 4	The C & D-Brain In Your Marriage	72
Chapter 5	Racism I	98
Chapter 6	Denial & Lying	103
Chapter 7	Abortion, Privacy, & The Constitution	112
Chapter 8	Personal Conduct	120
Chapter 9	Drugs, Alcohol, & Cigarettes	125
Chapter 10	Guns and Butter	130

Book II – Economics and Market Fences

Chapter 1	Deer Meat, Money	134
Chapter 2	Collectivism vs. Competition	140
Chapter 3	Combined Laws of Thermodynamics	146
Chapter 4	Effective Anti-Trust, The Holy Grail	165
Chapter 5	Anti-T Principles, We Can Live With	189
Chapter 6	What If Something Bad, Happens?	200
Chapter 7	Privacy Law & Lawyers	205
Chapter 8	Japan, World Unity, the Anti-Christ	211
Chapter 9	About War	225
Chapter 10	Racism, Convicts, and Immigrants	235

Book III – God is Love & The Unified Field Theory

Chapter 1	There is a God	250
Chapter 2	The Unified Field Theory	255

Appendix I – More on Hawking	289
Appendix II – A Sectional City & Moving Sidewalk	293
Appendix III – The Itch	295
Appendix IV -- The Energy Challenge	302
Appendix V – Lexicon	306
Predictions and Recommendation	311
Conjecture and Angels	313

PRELUDE

The Revelation

In the summer of 2004, I experienced what I call the Revelation. The main ideas presented in this book popped into my head, fully formed, *in one night* as I was sleeping, in the form of a multi-dimensional memory without words or letters, containing only ideas and images fused together.

It was quite unusual and very beautiful. Think of a blueprint composed of images: images that interact with each other in ways that physically demonstrate mathematical and logical relationships. This blueprint, which came in three distinct parts, depicted a multi-dimensional space in which my mind could wander, full of understanding displayed as ideas, equations, and images — but no words.

After that night, whenever I think about certain questions, a distinct part of the Revelation becomes visible. That part is a vision or image of something happening, like gas molecules in a cloud heating up. The blueprint has linkages across biological, mathematical, physical, and spiritual understandings, including how the human mind evolved here on Earth.

It is now the spring of 2009, and what you hold in your hands is my attempt over the past five years to put much of what appeared in that Revelation into words. I have not done justice to the Revelation with this book. I am ashamed of my lack of ability to translate these images and abstract ideas into an understandable combination of words. In fact, in several places in the book, I've had to invent words to try to convey a concept. I've included a Lexicon in Appendix V for that reason. My first efforts were so unintelligible, even my friends couldn't read them. In fact, I re-wrote the book over 20 times before showing it to a friend who did find it comprehensible. That friend was Dr. Mark Richards. At that point, Dr. Richards stepped in and has helped tremendously with the translation of these ideas, although even he and I needed five more re-writes before *we* were satisfied. To him and his wonderful family, I am eternally grateful. I hope that when the ideas presented here resonate with events and ideas experienced in your own life, you will begin to see the more complete picture of these concepts for yourself.

I am describing a vision of the world that can only be understood after you throw all your pre-existing mental constructs and prejudices aside and embrace the vision, even if momentarily. Once you do that, you will clearly see the world in a more understandable and much less confusing manner. After observing enough events explained by what you will learn in this book, you may feel confident using this interpretation of life full time.

The ideas in the book are highly unconventional, but once you see their truths, you'll notice them everywhere. I hope this translation of the Revelation, brought to you by Dr. Richards and myself, will enrich your life.

The Book is Presented in Three Main Parts

The first section, or Book I, is the *"Peace Machine"* Hypothesis. It's the story of how man's mind and brain got here, from an evolutionary standpoint. It explains how our development as a social species was directly affected by that evolving brain structure, something still true today. This important section presents a few different, rather novel ideas that explain how Darwinism shapes our society, our laws, and even our customs. I explore how Darwinism affects our personal and societal relationships, and, finally, why science and spiritual belief, which appear to be in conflict, are not opposed at all.

Book II is the explanation of the economic theory contained within the second part of the Revelation. This new explanation of economics, entitled "Market-Fences" builds upon the framework of our new understanding of the human brain as it relates to the *Peace Machine* Hypothesis. Briefly, the human species' novel brain structure affects our output of goods and services, and ultimately the distribution systems that transfer wealth inside human society.

I have resorted to an analogy to express this concept within this section of the book. The production and distribution of goods and wealth resembles a gas cloud. Humans and assets are like molecules in the cloud. They transfer heat (money) from molecule to molecule (person to person, asset to asset) depending on the productivity multiplier sitting atop each molecule (person/asset). Our existing legal and financial systems don't take full advantage of these principles, which is a big limitation on human society. Although the ideas expounded don't appear sexy or important on the surface, the information contained in this section is essential for each man and woman to understand if we are to prosper instead of perish as a species. The antitrust and Market-Fences part, in particular, was given great weight/specificity in the Revelation.

The third section is the last and the most important of the three: **God is Love**.

This concept is inseparable from the Unified Field Theory, a theory that explains the interconnection of God, physics, and the world we live in right down to the atom. The Revelation came with a lot of data, all interrelated and beautifully linked together: one principle supporting another, flowing from mankind's early development to the physics of the universe and then back into focusing on mankind and his relationship with God.

To The Atheists

If I partially lost the atheists here, I understand, because for many years I was one. But stay open-minded, and your opinion may change. Here is a little test. Stand in front of a blackboard with chalk in hand. Pretend that your professor has asked you to write a formula mathematically expressing the universe. The only information he has given you is that the universe you are to describe has a God. Work on that awhile. You will quickly realize that you cannot mathematically express this situation without including God in that formula. If there is a God in the universe, we must account for Him/Her/It: in our calculations, in our daily life planning, and in our science--all of our science-- as well as our spirituality.

Why can't a more complex level of consciousness exist above us on the evolutionary ladder, a consciousness of which we are ignorant and unaware, just as the ant is unaware of the level of human comprehension? Humans can clearly see hundreds of species below us on the evolutionary ladder, all lacking comprehension of a more complex rung above them, each possibly assuming it is the top rung — exactly as we do — and all of them wrong.

Why? Because man is in denial, about being a killer, about being a liar, and about the other choice God offers him. He is a predator capable of destroying every species on this planet, including his own. But he can be something far greater than that through the exertion of free will and the use of the *Peace Machine* section of his brain. Essentially, man can use effort, concentration, or —to put it another way--a mental nexus to realize there is a choice that comes not just from programmed, patterned, or instinctive behavior. However, that requires understanding both sides of the equations we are using to define the universe — and that's what I hope to do with this book.

About Me

Before I conclude this preface, I must confess: I am not a "good" man. When I was married, I cheated on my wife. I left her and two wonderful children through divorce. I lied practically every day of my life, and I hurt a lot of people along the way. I cheated virtually every system I came across in one way or the other, all my life.

Looking at me from the surface, you would not observe any of this. I appeared happy to help others; in fact I was highly successful in the eyes of society. In reality, looking back after the Revelation, I've concluded that I helped others only when it suited my purposes, made me look good, or otherwise directly benefited me. And, as mentioned, I was a complete atheist. I could, and did, lecture for hours about how stupid people were for believing in a fairy tale about historical characters and "miracles" in the past. Furthermore, I took delight in exposing the religious stories, contradictions, fake miracles, money shakedowns, and holy wars, using my skill as a speaker to rip apart believers with that information.

As this book will point out, the world I was looking at through those eyes was, and is, real. That is the irony. I was using a behavioral pattern successful for survival, and I had an existential honor system that I obeyed--but not for the right reasons or the right motivations, and therefore I was blind to real truth. I was living completely in denial.

You may be wondering: If some higher order of consciousness does exist, why would it (God) put a revelation into the head of a self-described loser? I do not know the answer, other than to say that He must have known I would write it down. In every other way; I am undeserving. That is not false humility; it is the truth. It has been an all-consuming struggle to interpret what's in the Revelation and translate it into words. If you meet me, you will be disappointed with what you find. Take these ideas, and then forget me.

I talk about the ideas in this book as if they are my own, because they flow out of my brain. But I am not that smart. I would have needed three or four college degrees to make the linkages across disciplines that appear in this book — and I don't even have one degree. Until the Revelation, I had no idea of the interrelationships of the physical, spiritual, and emotional worlds, nor any idea of the complexity of the threads that run through everything. How else could I have discovered that every black hole at the center of every galaxy is part of the same black hole? (Unified Field Theory Explained. Book III, chapter 2.)

Prelude: Conclusion

Don't be discouraged by the early chapters if you're an optimist, and don't be put off by the ending if you're a pessimist. Each of the truths revealed in this book requires all the others. They are forever interconnected. The ideas are presented in a specific order, as laid out in the Revelation.

I believe that everyone can see the Revelation's core truths. If you drop your intellectual conceits, ideas, ego, and pride, the understanding will come easily. But if you insist on clinging to your current beliefs, this book also mentions the scientific basis for these truths. It is science that you can research and read about on your own. The main reason we usually don't see God's message to us is that our own thoughts, beliefs, problems, ideas, and convictions are in the way. You have to stop shouting at yourself before it is quiet enough to hear God whispering to you.

I set this book's price above that of the average hardcover because even though people are thirsty for knowledge, the human brain perversely distorts value and cost. If this book were inexpensive or, worse, free, the reader would simply place that value on it: nearly worthless. If the Earth charged us for every barrel of oil we pumped out of her, we probably would have acted more rationally and not only used less oil but developed reliable and affordable alternatives along the way instead of having burned up half of the world's known supply.

We spent billions upon billions, back when that was big money, to go to a cold rock in space orbiting our planet, while the magnificent machine that every single one of us carries around between our ears that drives our entire existence has barely been researched. Read this book with an open mind, and I promise that many of the mysteries in your life will finally make sense. For perhaps the first time, you will have the tools to do great things and be happy in your life going forward.

You want understanding that begins with your everyday life and stretches past the cosmos to the heavens? I've written this for you. Make a commitment to give your brain a chance. Read this book carefully, even if it takes you a month, and then read it again. Think of it as the first operation manual for the human brain. That's what this is all about in the end: your brain and the truth. If you navigate these first few chapters, which I admit are somewhat challenging, I promise you the power to find the truth: in your life, your heart, and the rest of the world.

From Dr. Richards, the Translator:

To understand the true nature of the human mind is the key to understanding all that humans created. With deep understanding, we can restructure our personal, societal, and national systems/actions to mutual benefit.

To understand the creation of the universe is the key to understanding why we are here, how we got here, and our maker's prime directive to us. If we fail to achieve either understanding, we are as doomed a species as the powerful dinosaurs before us. This work was motivated by love in the hope that we may continue to evolve toward a finer creation.

Introduction

The *"Peace Machine"* Revealed

This book is a collection of ideas. Some of them you will recognize, and some are very different from what you've ever read or heard. The early parts of the book describe a system I call the *Peace Machine*. This system exists inside your brain and is the manner by which you make and execute every important decision in your life. It begins with the premise that you have a second brain, *and* a second mind, contained within the physical organ we currently think of as the brain.

> The *Peace Machine* process is derived from the human brain's unique architecture and the resultant routing of each individual's knowledge and memory flow. It is our unique architectural brain design that exerts control over human behavior to create a social brain (in essence, a *Peace Machine*), which allows predator/killer humans to live together in groups without the bloodshed that would negate the advantages gained.

That's right: This won't be hard to learn, because the *Peace Machine* is already in your head; you use it every day and don't even know it.

Most people would deny that their conscious brain shares power in its decision-making process with a hidden, more powerful set of instinctive instructions and impulses. But that is exactly what happens on two levels, physical and metaphysical. Scientists call what we consider our "advanced, logical, and thinking" brain our cortical brain. It contains the evolutionary adaptations described in this book to communicate with our older, instinctive, pre-programmed brain, which is physically located beneath the cortical brain. This setup is the evolutionary adaptation/solution I have labeled the *Peace Machine*.

Other scientists and authors are on to this idea, but they tend to see just pieces of the whole, not the universal whole. I will name a few of their books, and I strongly suggest that, after you read this book, you read theirs to get a different view of the same concept from a more traditional angle. Actually, considering how "far out" some of these authors are considered by mainstream science, it's humorous to refer to them as traditional.

The first is Harvard professor Stephen Pinker's *The Blank Slate*, a very important work. Mr. Pinker also wrote a jacket quote for Richard Dawkins, author of *The God Delusion*, another piece of recommended reading.

The Mind & The Brain, by Jeffrey M. Schwartz, M.D., and Sharon Begley, is a fascinating fusion of some of the ideas presented in this book from an approach so strait-laced it would be comical, except that the authors never pull the trigger and proclaim the dots connected. But theirs is a really great book. If you can't finish this one, at least read that one.

Also, Malcolm Gladwell's *Blink*, anything by K.C. Cole, and *Practical Ethics* by Peter Singer. All of those are brilliant in their own way. All surpass my modest effort here. But what I lack in academic brilliance and credentials, I hope you'll find compensated for with *relevancy*. I could draw a thousand parallels and a thousand distinctions between my work and that of the authors listed above. But I won't bore you. The works speak for themselves. Read this, read their titles, and then make up your own mind. Frankly, when I discovered there were other people writing about this, I was thrilled. It meant I wasn't completely crazy.

The most reasonable explanation for the lack of useful and relevant information available elsewhere on the topics in this book is the major role that denial plays in the history and behavior of humans. That denial role, due to how central lying and denial are to the *Peace Machine Process*, is much larger than currently understood. History textbooks are a partial illustration of this. Each race and/or nation writes history from its own perspective.

Each country's authors preen over perceived accomplishments and grieve over unfortunate events such as lost wars. Only who won wars is fairly recorded, and even that is bungled now and then. Some accounts of the same war or unfortunate event differ *so much* from race to race that it touches off more unfortunate events or wars.

The rewriting of history, exalting the heroic nature of man (often with a victor's account of what in reality was his savage behavior), has been happening for thousands of years. It is best in evidence when reviewing the history of China, Japan and Korea.

It occurred everywhere else as well; it is just that the Asians kept more complete written records. Even today, war films, books and histories present one side as divinely heroic and the other side as justifiably machine-gunned to death. The more savagely, it seems, the better.

The descriptions of that wasteful behavior are then re-transcribed over and over for future humans to learn. Our myriad collective histories, instead of telling the literal truth, describe our past wars rather gloriously, as though they were coordinated, grand, colorful and important. And it's not just generals, church leaders and politicians; scientists can be just as bad on this topic in their own way.

An intellectually honest review of history and science reveals that we really aren't so different/better/divinely inspired when compared to other higher-level animals. Our inherent-superiority bias exhibits itself differently in the scientist from the way it does in the spiritual or political leader. However, the nexus for it in the brain is the same.

Why all this collective denial? The truth is, our ancestors basically fought for our species survival on this planet, side by side against superior predators, for countless generations until our tool-making skills modestly elevated us above the other predators. That early skill with tools was an extension of a previous evolutionary adaptation, the brain adaptation designed to promote human socialization.

It is quite an irony that early man used our new technological advantage (tool-making) to make war on our fellow humans from the first moment we could and with everything we had at our disposal. We fought race to race, tribe to tribe, religion to religion, nation to nation, person to person for almost all our recorded and researched past, restrained only by the *Peace Machine*, the evolutionarily new social brain that allowed groups of human families to grow into larger social structures.

Despite the ever-growing social brain, the "Europeans" fought against each other so ferociously that they absorbed every spare resource. They lost their way as a species for 1,500 years (the Dark Ages). During the 1300s, one-third of all the humans from Iceland to Italy died off. Knowledge once known and shared disappeared from collective archives, consciousness and memory.

People got poorer and stayed poorer for hundreds and hundreds of years, not that long ago. Without some change to our social and economic systems, the human race faces another die-off soon.

Might Is Right, or Deer Meat Equals Money

Economists call the sum of what we produce and make the gross domestic product, or GDP. The phrase is "fancy talk" for our entire work product: all the stuff we do, whether it's to cut someone's hair, build a car or grow vegetables. It can also be expressed as wealth, because it is the change in GDP from year to year that determines whether the economy is growing or not.

> The population in a community or nation with a growing economy, despite that population's position relative to other communities or nations, "feels" wealthier while its economy is expanding. Expanding economies generally make people happy.

In pre-World War II Germany, people's survival brains were happy with Hitler because he had wiped out the bad economic times of the Twenties and early Thirties by putting everybody to work in factories.

The Chinese people's survival brains today are happy with their government, although it wasn't selected by them, because their GDP is rising rapidly. As this book will explain in great detail, actions that put more money in the bank, more food on the table, more "deer meat in the cave," are always interpreted by certain parts of the brain in a positive way.

Incidentally, it may be worth noting that the Chinese government today is accomplishing this juggernaut of political fusion in the *exact* same way Hitler did: putting the people to work in factories. Whether the Communist Chinese leadership's level of intolerance to all other thought is as rigid as the WWII Germans' remains to be seen. To pretend that populations don't think and react this way, is naïve and dangerous. To not learn from this is to continue your misunderstanding of your country's history (and your own personal motivations), and that, collectively, serves as an invitation to war. It is a terrible miscalculation that may land the United States in a war with China, a war the United States cannot win.

On the flip side, people are generally much more pessimistic when the economy is not doing well. Today, an economist would term two quarters of GDP contraction a recession. Imagine if the GDP, that key core measurement, were declining per capita for humans all over the globe for 6,000 quarters in a row (1,500 years), as it did during the Dark Ages.

Democracy Is Dependent on Deer Meat

You may think freedom is important to you, but in reality it's not as important as food. Power-sharing humans (democracy) will not be satisfied with any government they elect or bring to power, regardless of its initial standing with the people, if it doesn't produce economic growth. Of course, non-power-sharing populations can't make a peaceful change; they live under tyrants who suppress, torture, rape, financially pillage, and brainwash their subjects.

Times have been lean for humans for a long time. Extremely lean environmental conditions are what shaped how our ancestral human brain developed its behavioral patterning. The programming rules that mainly govern our present behavior were selected by the subsistence environment our early human ancestors faced; not our modern, indulgent, environment.

Estimates exist on worldwide GDP in something called "adjusted" dollars for as far back in time as humans have records. The estimates presented here come from J. Bradford De Long, in the Economics Department at the University of California, Berkeley. According to Mr. De Long, the GDP peaked before the Dark Ages in 500 BC at $130 (per capita, annually) and then declined to its bottom at just under $100 from between the time of Christ's death to the year 200. It gradually rose again to $130 in 1000 AD, and then fell again for 500 years before permanently recovering above $130 in 1500 AD. Substantial, lasting GDP increases were not reliably produced until after the invention of the printing press in 1422.

The beginning of humans' new age of productivity and prosperity was tied to the widespread distribution of reading and writing. It is that process that matures the cortical brain, getting it physically and mentally developed enough to operate at a productivity/comprehension level sufficient to challenge the older predator-brain for control.

This new age did not begin or coincide with the discovery and widespread use of petroleum, as theorized now by many modern doomsayers, who credit our surge in productivity with the use of oil. These are the same doomsayers who predict the world's end and/or economic chaos when the oil runs out. I will touch on that in the Economic section of the book. Our increasing GDP is not solely from consuming fossil fuels. However, the doomsayers may turn out to be right -- not because we used too much oil, but because we failed to plan adequately for the end of the supply.

After 1422, it was "off to the races." Using this same method of counting GDP, worldwide GDP today is $6,500 per *person*, for over 6 billion people. That is a remarkable achievement. It was accomplished in only 500 years, because we could reliably spread knowledge about the different technological solutions available to us.
Reading and writing elevate each human into a more productive organism

through this brain process. The written word's widespread use also pushes productivity increases into the greater population, increasing the output of the entire system. Biologists will tell you the brain cannot change that fast. They are wrong.

Our brain's ability to change a decision, or transform a powerful urge into a physical capability through the constant evolution of the mind, neuron by neuron, is ever-present. This is called brain plasticity.

Humans are Predators

On the negative side, we used our newly found knowledge base and technological advantages to find, extract, and burn half of all the petroleum in the world.

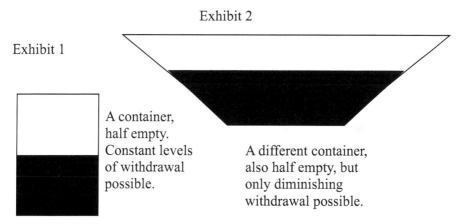

Exhibit 1

Exhibit 2

A container, half empty. Constant levels of withdrawal possible.

A different container, also half empty, but only diminishing withdrawal possible.

We did that with no regard for the day we wouldn't have oil. Look at the two pictures above. The first one represents how we mentally picture something that is half-gone. Now look at the second exhibit; it represents our actual situation, worldwide, with the oil supply for the next 30 years. The human race will be constantly trying to suck the remaining half of the oil out of a narrower and narrower supply tank.

This is collective denial at its most spectacular. Humans pumped, refined, and used the energy from the top half of the "tank" to spread cockroach-like from one previously unpolluted, un-mined, un-radiated, un-poisoned, unburned, un-bombed corner of the globe to the next temporarily un-ruined spot, *never* looking past the supply collapse. We prefer collective denial.
It's no accident that the first significant thing humans did with the power

derived from the splitting of the atom was to create atom bombs and drop them on two cities full of other humans. No moral judgment here, just an observation. In fact, later in the book I will show how the atomic bombs saved more human lives than any other invention of the 20th century.

All of this unhappy oil supply news is only part of the tale. To see the whole story, we have to start seeing the enlightened deceptions (rationalizations) within our individual brains for what they are: lies, conceits, or necessary adaptations, depending upon your point of observation.

Picture yourself. What kind of a person are you? Fair? Hard-working? Incapable of murder? Now forget about that image, and try to create an image of what all of us humans are collectively. Is it still just a bunch of fair, hard-working people incapable of murder? Now compare the combined image of all the people in the world with your self-portrait. The second image is much closer to the truth about what you are than the first image you created.

We must understand denial and our ingrained reaction to the truth, especially negative truth, in a whole new way if we are to understand the *Peace Machine*. To comprehend the *Peace Machine* at a fundamental level requires a new understanding of the human brain: how it developed, what it does, what it *admits* to itself that it does, and what it admits to other brains.

The understanding of which parts of human behavior still promote our survival and which parts served us well in the past but are no longer useful explains not only our current individual and collective behavior, but also the underpinnings of our economic systems and the creation of the universe.

Humans are Lying and Enlightened Predators

Our laws in the United States, the laws of other societies, the laws of economics, and even the laws of the universe are all related and are interdependent. Don't try to fit the ideas presented in this book into the usual drawers in your mind. They won't fit. Many things we view as either "right" or "wrong" are actually different parts of the same organism, and as such are dependent upon each other. In those cases, can either side actually be fully right or wrong?

The *Peace Machine* part of the human brain was a creation of those principles. That's why learning about how it works can open your eyes to the invisible threads running through everything around us.

That may sound far-fetched, but when you finish this book you'll be able to see the interlocking threads for yourself. Then you can struggle with the really tough part: what can be changed, what's next in our evolution, and what undesirable traits are we stuck with as tool-making humans that only God and love can help with.

> It is not just our traditions, morals and customs that are relativistic; our actual *logic* and legal systems are relativistic. That wouldn't be possible without individual and mass denial.

Two of the world's greatest documents, the Declaration of Independence, penned by Thomas Jefferson, and the U.S. Constitution, are perfect examples of our mass denial. Despite the existence of the Bill of Rights, those rights did not cover the majority of people living in the colonies during the time both documents were penned. That contradiction (especially in relation to the slaves) tragically undercut many of the principles enshrined in the Constitution. It eventually led to the Civil War, costing the United States a total of 600,000 young, productive men out of a population of only 30 million. All of that death and destruction occurred less than 100 years after the Constitution was ratified.

To give you perspective on that number, we would have to lose 6 million dead young men and women to a war today for it to be numerically equivalent. Six million dead! One *hundred times* U.S. losses in the Vietnam War. Lincoln delivered us from that slaughter and gave that sacrifice of life meaning by making the inalienable rights outlined in the Constitution's Bill of Rights mean something for the first time.

None of the *seemingly* political statements to be found throughout this book are meant to be political. They *are* intended to shock and force you to question everything. Make your skepticism official; if you don't, you'll be misled. The Devil's Advocate was an official post in Rome. The Bishop or Cardinal assigned to research a proposed saint, acting as the Devil's Advocate, would gather the necessary evidence and testimony needed to discredit the miracles attributed to the proposed saint. It was thought this would balance the process and lend it credibility. During Pope John Paul the second's term, he abolished that centuries-old office, dispensing with it altogether by papal order. Perhaps he decided it wasn't politically correct to ask if a saint could also be a sinner. As a result, John Paul II oversaw the canonization of more saints during his term than all the other popes *combined*!

Always play the Devil's Advocate: as glorious as the U.S. Constitution is, try to see it from the point of view of a slave, or a woman, or a white man with no property. If you can do that, you should be able to see the contradictions in your own brain, in your own conduct and in society's conduct. We need skepticism to overcome our own natural inclination to change reality around inside our brain, to fit what we want to believe instead of what's true. Be skeptical of what you currently believe -- the most.

On Mr. Jefferson

I think Thomas Jefferson was a great man. I believe he meant every one of those words he wrote in the Declaration of Independence. I also believe he owned slaves. I believe he loved and fathered children with a slave, a woman he owned. His own children by their slave mother were *not* free until he set them free after his death in his will. Still, Mr. Jefferson did not free Sally Hemings in that will, *nor did he set free the remainder of his 130 slaves in his will*. They were sold, along with his land, to settle his debts, and she would have been sold, too, if Jefferson's daughter had not interceded to make provisions for her to live out the remainder of her life un-indentured.

Most Americans, when asked to name their favorite presidents, list Mr. Jefferson along with Lincoln, Washington and an interesting group of others including John F. Kennedy, Theodore Roosevelt and Franklin D. Roosevelt. To deny the truth about Jefferson is to deny the basic construction of the brain itself, which made all of those contradictions possible in the same man. Today, if you read about the above behavior and didn't know it was by Thomas Jefferson, your cortical brain would declare him an immoral hypocrite. In reality, his *Peace Machine* process got two things done: It kept him alive while allowing his intellectual brain to imagine a better world. Your brain is doing the same thing right now. Are you a hypocrite?

Try New Observation Points

With that frame of reference, we can more evenly consider the U.S. Constitution's greatness next to its flaws. We can more easily see other points of view. This shift in perspective allows a *new* observation point, which is necessary if we truly want to see the *Peace Machine* exposed: its weaknesses and strengths revealed.

This book may appear disjointed: discussing Darwinism here, racism there, physics, and then more racism -- all disparate subjects. But don't be linear in your thinking; instead, try to think how these topics might overlay one atop the other. To begin this journey is to accept that these threads, contradictions, and denial mechanisms/peace machines can co-exist in good faith inside each human, a human capable of the greatest enlightenment while simultaneously being driven and controlled at the deepest level by remorseless, killer, super-predator programming.

There is a significant reason why this brain split hasn't been previously revealed in this manner: persecution. Historically speaking, people who speak new truths about the cosmos and the affairs of man are rarely recognized for it and are more often than not criticized, attacked, and even killed. The inciting factor doesn't have to be the exposition of some challenging truth either; humans will kill for almost any reason.

For that reason and others, this book is not written as a scientific text, as *Origin of Species* was. That magnificent book by an incredible and compassionate genius was, alas, too scientifically detailed and too supported by fact. It was so dense it was never destined for mass consumption. Much of its beauty, therefore, remains unknown to this day. Almost every single criticism leveled at Darwin and the theory of evolution was answered to complete satisfaction in Darwin's writings themselves. But because not all of his fans and few of his critics bothered to read his work completely, much misunderstanding about Darwin abounds.

This book is meant to do something altogether different. It's *meant* to be analogy driven, illuminating a whole, not dissecting every little detail. I'm not trying to convince you of any one theory or idea. I am trying to teach you to see your brain for what it is. Understand your brain biologically: what patterns it runs, how it got here, what it's thinking deep down inside, what it's thinking up on top, and how those two are not only different and conflicting, but also how they are constantly interacting and relate to each other.

Once you can see your brain's relation to the rest of the universe, you can see the patterns, the history, the conflicts, the ideas and theories from Darwin to Dilbert, and judge them from a different *point of observation*. That is this book's purpose: to help you understand your individual *Peace Machine* and master its use.

You Have a Choice

That motivation and purpose separates me from the previously mentioned authors who are technically and educationally my superiors. They, in my opinion, are dissecting rather than assembling. That obscures the fundamental truth: **We all have a choice.** Consider that last point over and over again as you read this book and are given examples of how the *Peace Machine* affects society, race, law, war, health care, democracy, sex, etc…and just as important, where it doesn't. I am not taking a stand directly on any of those topics. They are only provided to you, the reader, to illustrate a *Peace Machine* point (a brain design point), a Unified Field Theory point, or even a spiritual point. I am not taking sides in any political, religious or social controversy, despite the connotations *you the reader attach to it.*

> A person who understands the *Peace Machine* process can make up his or her own mind, on every imaginable decision and principle, without any advice, coaching, or guidance from me. Understanding leads to a pure exercise of free will, as it should.

I have relative anonymity in my life at the moment, and I want it to stay that way. But if you meet me along the way and we get to know one another, I'll talk your ear off about the *Peace Machine* — if you'll permit it.

> I want you to learn about the *Peace Machine* because it's in your head right now, making decisions for you. Most of those decisions it can handle without your conscious help. But, in some areas you could do a lot better if you understood the basis upon which your "brains" are acting. Make both brains work for you.

Finally

A new point of view requires understanding what wasn't accurate about the old one. Hence the order of the chapters in this book, focusing first on the evolutionary biology and science that got man to where he is today. Next is a discussion of the physics of economics, and then an understanding of the Unified Field Theory, God, and how physics connects it all together in "our side" of the universe. The juxtaposition of science, politics, and spirituality is purposeful. They are connected, and we can't understand any of them fully until we acknowledge that connection.

BOOK I

The "*Peace Machine*" Hypothesis

Book I, Chapter 1

Let Down Your Guard

I need you to do something very difficult, so that you can understand what you are about to read. I need you to read the following pages without mentally arguing with every little thing you currently perceive to be untrue. The reward will be an understanding of something totally foreign to you: *The Peace Machine*. But to get there, you must begin with the premise that what you read here will disagree on many levels with what you've been taught all your life. If it helps, think of this as a fictional story. You must willingly suspend your disbelief for the duration of our time together in order for the knowledge that follows to add up in a coherent and useful manner.

Before everyone knew the Earth was round, everyone knew the Earth was flat. They even thought it had an *edge*. That wasn't so long ago. The simple truth is that you are equally ignorant about a number of important subjects. Open your mind to the idea that you just aren't as knowledgeable as you think you are. Additionally, your conscious mind hasn't made nearly as many decisions as you believe it has about your life. Even more surprising, many of the decisions you made were not for the reasons upon which you think you based the decision-making process.

Assuming that the previous statements are true while reading this book will prepare you for the real explanations of the truly mystifying things in your life. This includes your professional life, your family life, your sex life, and your position in the universe.

> The answer is buried within your own mind in the form of a second consciousness, literally helping you and hurting you in a thousand big and small ways each day. This second consciousness is overwhelmingly powerful, and is the reason we have survived as a species. It has no language, no distinct voice to be heard. Yet, it is the force behind your existence.

Humans are close-minded, suspicious, and judgmental. To understand this book will require you to keep an open mind. Abandoning resistance to a new perspective demands great mental effort. For that reason alone, some will not finish this book. It will simply be too much work.

Ideas, principles, and knowledge accumulate behind certain barriers, like water behind a dam. Once humans accept (rightly or wrongly) something as true, they build upon that perceived truth, brick by brick, layer upon layer. Sailors tried to keep the shoreline in sight in a flat world because they knew from experience that it was dangerous to go farther out to sea. Dangers did increase far out to sea, but their understanding of those dangers was false:

> *"The edge has to be somewhere out there at sea, since no one has found the edge on dry land."*

This sounds similar to the philosophy arguments we hear today, only without the word "edge." If I had said to an ignorant fisherman a thousand years ago:

> *"Doesn't it make sense that since we haven't found an edge on land, there probably isn't one at sea, either?"*

The ignorant fisherman would have hit me on the head with a stone and hidden the body under leaves and branches, while muttering:

> *"The less talk about going farther out to sea, the better."*

One tiny miscalculation (lie/denial or ignorance) at the root of a complex system can create a lot of havoc in the final result. For a modern case in point, reflect on Microsoft's experiences after each launch of a new operating system. This very real struggle between man and science actually began as the struggle between religion and science, with Galileo cast as the central figure in *the* drama of the centuries.

Galileo's challenge still shakes the ground today. His declaration that science was true – regardless of religious teachings – was so important to the development of modern mankind, that even the Pope appears just in a supporting role. Yet even today, the chances that we think like the Pope are still greater than the chances that we think like Galileo. Why is this? In reaching the decision to threaten Galileo with torture and condemn him to house arrest for the remainder of his life, so as to secure a recantation of Galileo's revolutionary theory that the Earth orbits the sun, the Pope had to overlook several significant contradictions:

…The Pope had a long-standing personal friendship with Galileo.

…Galileo had a deeply held faith in the Catholic Church and God.

…Galileo's daughter was a nun.
…Galileo's theories were based on the work of Copernicus, a Catholic monk,

published by the Catholic Church.

...And finally, Galileo's observations were backed up with evidence from a telescope, which at the time was a new scientific instrument. But none of this mattered to the Pope, because Galileo's conclusions were not consistent with the false beliefs of the time. There was just too much water (lies, denial, mistakes, unknowns, guesses) behind the dam demanding an Earth-centered universe. We laugh at the Pope for the futility of his struggle with Galileo, not realizing how much our actions are just like the Pope's. I want you to let go of the pre-conceptions that block your ability to see the true picture. For that to work, you must open your mind to the possibility that the entire way you look at things today may be false. This will be easier to do if you remember that our species' tenacious hold on false beliefs in the face of overwhelming evidence to the contrary had an evolutionary purpose: so you could live with other humans.

Most of the important decisions and actions taken during your life are for reasons different from what you consciously believe them to be.

How could that be?

How could the Earth be round?

People would fall off!

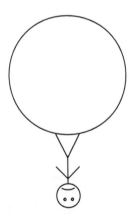

Unless you know about gravity; then it makes sense and appears obvious. Soon, the construction and function of the human mind/brain as it evolved over eons (our *Peace Machine*) will be explained. Then, the way you look at things today will seem as outdated and ridiculous as the idea that there is an edge to the ocean or that the Sun orbits the Earth.

Book I, Chapter 2

Super Darwinism

So what do God, Darwinism, The Unified Field Theory, Sex, and an obese America all have in common? The *Peace Machine* of course. We'll start with Darwinism. Since all of this material is closely interrelated, I'm obliged to start in the middle. When you finish this book, *I advise reading it again within 60 days*. Much of what you missed the first time or did not understand will make sense the second time.

Let me start by stating that evolution is a fact. Evolution doesn't just mean that man is related to monkeys. It means that **all** the creatures here on the Earth are related to each other. The concept is actually bigger than that. EVERYTHING is interrelated and exists as a result of evolution of the cosmos, which has as only one of its great-grandchildren, biologic evolution.

Some people think the fossil record is somehow false, and that dinosaurs weren't here 65 million years ago. That belief *requires* that the fossils and their existence in the rock beneath our feet are a pre-meditated con job by God (or equally powerful entity). Why would God do that? If God was capable of such a lie—why would we ever believe that God again? Let's assume that God doesn't mislead. The dinosaurs had a purpose. Maybe their purpose was a warning to humanity, literally written in stone:

> *"Look to the heavens before it is too late--before death and destruction come hither—again."*

In the very early years, perhaps Pre-Cambrian, about 3.8 billion years ago, single cell species (an early version of Diatoms?) living in a volcanic pond began to run out of food and/or water. While most all life in the pond died, two or more single cell creatures had a genetic variation that caused them to surrender their individual cell walls, link up, and share the remaining water. In this way, "Sex" was born.

Early sex probably didn't feel good, but survival did. Whatever abnormality in their simple genetic make-up that occurred to allow the joining of cells was a hugely successful adaptation. As a non-scientist, I can't plot out each twist and turn these simple organisms took from there. Use your imagination and look at the variety and abundance of organisms today. One can easily imagine the path biologic evolution took over the millennia to create such diversity. Modern day humans have no experience with our planet as glacier-covered, or dark after an asteroid strike. Consequently, we are in denial about how much of our planet's history was spent this way. We are equally ignorant of the powerful forces that shaped modern life's evolution.

We could learn much if we could communicate with the *millions* of species wiped out by the planet Earth during hundreds of extinction events such as ice ages. The irony of this is delicious—many humans are worried we are damaging the planet (true) and that she is this benevolent mother who will never hit back (untrue). In just the last 400,000 years alone, some of the most humanly hospitable time in our planet's history, glaciers and ice fields covered the globe top and bottom, down to New York for example, for 250,000 years. **The majority of that time period was an ice age.**

Not all the branches explored by the evolutionary tree are still viable. In fact, the majority of species that populated evolution's branches are extinct. Importantly, many of those failed variations of life contributed to the knowledge base shared by all the surviving species. As far as advancement and preservation of species inside evolution: *what not to do is at least as important as what works.*

As the core groups of especially well-adapted creatures and their parasites progressed along the evolutionary ladder, they continually added to the DNA information base they shared with each other during the reproduction process. These new creatures transmitted genetically programmed physical and behavioral information derived from successful ancestors, and their survival rate was enhanced. None of this evolutionary history/theory is new, until here.

> The idea that all these mutations must have been random and therefore were slow is naïve and wrong. A random explanation for these processes does not work mathematically, nor does it correlate with evolution's observed time lines.

The incorrect belief that random mutations alone are the principle force underlying evolution is still current scientific dogma. Oddly enough, the mathematical difficulty presented by this theory has become a target for many religious and extremist groups. They gleefully cite the mathematical proof of the need for a non-random mechanism in genetic mutations as evidence of God, and proof that evolutionary science is false.

Meanwhile, many scientists tenaciously cling to the random mutation nature of Darwinism because they find Darwinism believable, even provable in so many other areas. They think this kink will be worked out by science in time.

So, in a really strange twist, scientists are accepting of the random mutation theory despite mathematical proof to the contrary because they have Faith in their Darwinist beliefs.

Prior to the evolution of higher order creatures, the first of several types of feedback mechanisms that service the cell at the molecular level randomly developed a way to influence the changes (mutations) in a certain *direction*.

This *extra* feedback mechanism (random mutations also concurrently occur) is driven by environmental factors. Hence the mutations occurring are actually invoked/directed by the changes in the environment—thus increasing the organism's ability to exploit the new environment in a positive manner more quickly, more precisely, and through more than one adaptation at a time.

For example, certain reptiles' sex is determined by the temperature at which they are incubated while in the egg – a dramatic example of how environment can determine which genes are chosen to become active in an animal. In humans, celiac disease (an allergy to wheat) is a genetic disease that is triggered by external events such as severe stress before it becomes "unmasked" and active.

Even more interesting is what evolutionary benefits these environmentally induced gene activations produce. In the case of the lizards, the mechanism produces more fertile female animals during more hospitable times. As our knowledge increases, we will find an extra-ordinary number of genes that become active or inactive in response to the environment. In another example, on a more systemic level, an acid environment may unblock sections of DNA that are not normally transcribed in a neutral environment.

This process of "directed changes "utilizes much of that huge part of the DNA strand scientists thought was "junk." Consider the mud wasp, born after its parent wasps die. It nevertheless has a perfect knowledge of how to survive and perform the complex tasks of building a mud wasp tunnel and stocking it with food for next year's baby mud wasps. This is inherited memory. According to current science, you don't have this ability. In fact, most people will tell you they don't have this ability.

You are distantly related to that mud wasp and have that capability just like he does. In fact, you have inherited significantly more advanced capabilities that are as yet unrecognized by humans. We use the output of these abilities daily without conscious recognition. Evolution just provided humans with an alternative manner in which to use our inherited memory information, and for very good reason as we will see. Hopefully this book will help you identify and understand the potential and objectives of that part of your mind.

Evolution produces alternative strategies for using virtually every available resource, thereby assuring success somewhere. This evolutionary process, this natural pressure to consume by one process or another every available resource, is an example of a natural process that is tied to the laws of thermodynamics.

The Laws of Thermodynamics, we will soon see, rule the universe from end to end--although I don't want to get into that until the later chapters on Economics and The Unified Field Theory. You act in certain ways because the Universe you are derived from does so. You are from it and a part of it, and therefore will act according to its rules and laws.

You have inherited memory. But, you've been taught from birth by your parents and modern society that you don't. Humans are in denial about this. This particular lie is why you don't know about the *Peace Machine* in your own brain. You currently don't even know the truth and reality of how you make important decisions. You will by the end of this book!

Monkey Business

Tabula Rasa (Blank Slate) is the philosophy that we are born blank and our mind is written upon by experience, which in turn influences our decisions and morals. Mr. John Locke, who along with Rousseau heavily influenced the thinking of the United States' Founding Fathers, first proposed this idea. It is wrong, although not completely. It is one of the ideas you must be willing to reject, if only temporarily, so that you can comprehend this book.

One part of our brain does begin somewhat as a blank slate. Since the part reading this book lives in that part of the brain, this Tabula Rasa idea "feels" right to us. But, there is another reason it "feels" right, our strongly- evolved tool of denial. Read Harvard Professor Pinker for more de-bunking of the Blank Slate idea and the power of denial. His published works focus on this aspect of the mind in the type of detail I am purposely avoiding here. But, if this part of the subject fascinates you, please read Pinker after you finish this book.

Don't worry about the fact that you don't *think* you have any inherited memory right now. Remember gravity? A round planet made more sense after understanding it.

> Inherited memory will make more sense if you think of it as knowledge passed down into your brain from a million years of evolutionary trial and feedback. The data is readily accessible, just not directly by your conscious mind. Inherited memory is not only knowledge. It's morals; it's emotions, it's *most* of what you use every day.

I know what you are thinking: this is different from gravity—if I had inherited memory, wouldn't I know it? But, what if you had inherited memory stored in an area of the brain different from your conscious mind, an area into which your conscious mind isn't allowed? What if I told you that it is precisely because your mind split like this a million years ago that man became the most dominant species to have ever lived on Earth?

Millions of years ago there was an environmental change, most likely weather or other natural disaster, which forced a species of monkeys, and probably many other species as well, to alter eating habits or become extinct. These monkeys added meat to their diet. What is the significance of meat?

These monkeys were already social, as monkeys typically are. They groomed each other for insects--the beginning of "touch" love--and did other tasks like raising their young for an extended period of time. Since they started out as herbivores, they were naturally more social. Creatures that don't eat meat don't fight as much amongst themselves. The aggressive "killer instinct" is not as well developed in most herbivores. The unusual addition of meat to this social herbivore's diet provided a highly concentrated source of calories, nutrients, and proteins. This allowed even more time for social interaction, as less time was now needed to seek and consume sufficient amounts of calorie-poor food such as leaves and twigs.

Advanced nurturing was good for primates. Being protected in a group that encouraged nurturing allowed these monkeys to increase their young's learning period and push their brain capacity enough to abandon using *predominantly* inherited behavioral patterns. On a new diet that included meat, they had more calories to use and more time for social interaction and *creative learning*, as opposed to constant food gathering.

Advanced nurturing also led to increasing emotional and *communication* abilities. In humans, increased communication skills were uniquely coupled with extraordinary visual abilities.

In retrospect, this dominant visual brain characteristic was a very important evolutionary selection. The human brain's visual center allowed it to become more precognitive (not in the clairvoyant sense), more mathematical and premeditating. The abilities to communicate and anticipate in advance are what were coupled together in this "new" conscious mind. This began to really differentiate us as predators.

We're different from the other predators because we kill more efficiently than they can kill. The contribution of precognition from our visual center, communication abilities derived from advanced nurturing, and the denial or "blank slate" portion of our mind (*Peace Machine*) that allowed more peaceful co-habitation— all fused together to create the ancestors of modern man. These traits allowed us our ability to make tools, another evolutionary edge.

>Evolution produced advanced nurturing as one alternative for dealing with scarce resources. This in turn eventually led to a highly integrated and developed *Peace Machine.* All of these adaptations share a common nexus, or impetus: love. This potential choice has enormous scientific and spiritual significance.

The Eyes Have It

The early monkey brains, which evolved into early human brains, processed visual images as ideas. These ideas began as encryptions, to speed visual traffic in the brain. All brains of higher order creatures need this adaptation to operate efficiently enough to survive.

In reality, it is just a bandwidth problem. The brain cannot receive enough data down the optic nerve (and other neural pathways), interpret it, and then order a response in time to save the viewer from death by a fast blow or flying object. That is unless the object gets converted into ideas (smaller packages of data), which represent *the threatening image.*

Images to symbols, symbols to ideas, ideas to action--all this became possible inside the early human brain because the human *mind* evolved to prioritize sight (over smell for example). This is just one example presented in this book of the three or four adaptations that happened closely in time to produce the modern human brain. Hunting meat required increased killing of non-humans, hence our species' natural inclination, ambition, and skill at killing. The evolving monkeys needed a section of the brain walled off from the genetically/instinctively-programmed area. They needed brain capacity in an area where experientially learned knowledge could be stored and used during the creature's lifespan. This was the beginning of the enlargement of the C-brain, the education and social tool created by the D-brain, the core or base brain, the pre-programmed <u>D</u>arwinian brain.

Even today, the first things humans do once countries get wealthy enough, is begin eating more meat. It takes 600 calories of grain to feed to slaughter 100 calories of meat. While this is horribly wasteful behavior, humans immediately undertake this process anyway--because their brain tells them to do so as soon as it is economically feasible.

The Two Brains and their Poor Relations

From this point on, I will use the phrase D-brain, to refer to the larger, older, evolution-shaped part of the brain that controls base functions and emotions as well as thousands of complex programs inherited from generations of evolution. Why the D-brain? In honor of Charles Darwin, of course.

You are in denial about the existence of the D-brain because your consciousness mostly resides in the other layer, your C-brain, or Conscious brain.

The D-brain is 70 to 80 percent of the brain by volume and is located, not on the left or the right side, but is physically *beneath* the cerebral cortex, the upper most layer of the brain (the one containing most of the C-brain). A small part of the C-brain extends down into the D-brain, like a driveway that leads to a different house. You can also think of the D-brain as the deep brain.

The D-brain is very focused on maintaining a constant vigil against death, tempered only by a strong impulse to have sex and procreate. This *base layer* of our brain makes **most of the important** decisions in our lives *even though the C-brain is unaware of this fact.*

The D-brain is home to denial, love, hate, rage, racism, greed, satisfaction, and most importantly, evolution determined behavior patterning—the real secret to our current behaviors *and* past survival.

For those who find this an insult to man's enlightenment potential, consider the more amazing truth—that from such depravity, from such humble and primitive urges, your brain has found a way to comprehend and choose love, or in other words, attain enlightenment. As an atheist like Richard Dawkins would say, any other story lacks evidence.

The Conscious mind (the part reading this sentence) is contained in the smaller, weaker, "blank slate", tool making, calculating, and socializing brain that I have labeled the C-brain.

The C-brain is located physically on the topside of the brain, in the cerebral cortex and is only 20 to 30 percent of the actual brain mass; doctors call it the "Cortical Brain." This is the home to logic, calculation, reason, charity, and intellectual work. The D-brain considers charity, forgiveness, and displays of emotion other than anger or rage as weaknesses that threaten the survival of the creature.

This basic C- to D-brain contradiction is why most people display both sides of these behaviors during their lifetime, though we as individual humans don't classify people that way. For example: we generally view someone as either more or less forgiving, but we rarely calculate how that compares to their treatment of others. Think about it.

Most people aren't either unforgiving or forgiving. Most people are very forgiving of some people (mostly their children) and are generally unforgiving of others. This runs through all groups of people. Religious beliefs can alter our *perception* of this reality, but when studied scientifically as a group, Catholic priests would not come up any more forgiving than atheist liberals. One evolutionary group over time would out-live the other if it were any different.

What does make a difference is what layer of the brain is making the decision. If a religious woman, who is related to you by blood and older than you, gets to make the decision about whether or not *you are to be forgiven,* no matter the evidence or the crime you will probably be forgiven. That's D- [mom] and C- [religious] brain in agreement working on a decision for you.

Conversely, if a white adult male is on a jury judging a person of color who allegedly committed a violent crime against a white human, the accused will probably not be forgiven, or even be given the benefit of doubt, regardless of the evidence. That's the D-brain (racism) overruling the C-brain (denying any evidence that would create doubt). Naturally, the juror would have a logical reason for his conviction vote—as would the related woman with her innocent vote. However, the logical reasons given wouldn't be the true thought processes. The D-brain programming is the true reason. Naturally, the C-brain has its own denial mechanism in order to maintain a logic facade. The C-brain will create this facade when the unconscious D-brain overrules it, no matter how logically absurd the necessary reasoning.

> Love and lying/denial are the *only* two base traits represented in *both* layers of the brain.

Because of its complexity, I don't want to get into a long explanation of how human speech is affected by all this. Suffice it to say that in controlling the voice box, the C-brain makes words while the D-brain controls volume, tone, and inflection provided the speech has emotional importance. This is another example of a concurrent adaptation. To help grasp the concept incompletely: words and *their attachment* to meaning are the creation of the C-brain. However, the idea itself in the brain, the concept, as well as tone, volume, inflection, sarcasm, sexual invitation by purring or grunting: these are all parts of speech that *can* be controlled by the D-brain.

A person raised in Mexico until age 7 has an idea of "Mother" in the D-brain from birth. He forms the word "Mama" in the C-brain (in Spanish) as he learns to speak. If then moved to the U.S., this C-brain can learn *to think in English*, after learning and using English more--forming a new word attachment (Mom) in the C-brain to the old idea/image of mother that *never changes* in the D-brain.

> All modern humans share a basic C and D division of the brain. It is the interplay--the way these two sides work and fight for control of a person's actions--that is the essence of the *Peace Machine* process. Each human's unique C/D interplay determines how successful he or she will be. Remember, killing is a method, not a reward. Deer meat (money) is the reward.

This responsibility split is the heart of the *Peace Machine*. Understanding each brain and what each layer controls, as well as the relationship the two sides have with each other, is the key to understanding your own behavior and the behavior of all humans around you. The newer brain area, the C-brain, began as a tool of the larger, stronger D-brain. The C-brain developed to enhance survivability, since pure D-brains urges were likely to lead to constant inter-species killing and warfare, which undermines social cooperation for food and sex. The older, more hormonally influenced D-brain is in charge from birth. It then makes an individual decision in each young human's life about how much space to surrender to the conscious mind, the C-brain--the brain responsible for peace with other human brains.

Because of the order in which the two brains developed, the D-brain retains ultimate control. However, the C-brain's unprecedented success as a hunter has upset the brain balance hierarchy in modern man as seen from 3 different angles:

A. Mankind's relationship with the other Earthly species,
B. Man's relationship with woman, and
C. Man's group dynamics with other groups of humans.

If we fail to apply C-brain adaptations in any of these three areas, we will suffer significant population declines and possibly extinction. Common words for these problems include: A. Deforestation and/or over-fishing, B. Divorce/Family Dysfunction, and C. War.

While most humans adhere to either a nature or nurture philosophy, the truth is a compromise between the two. Each child's brain moves incrementally from the starting mix of C- and D-brain allocation inherited from its parents.

This can go either way depending upon environmental influences. Humans that live in harsh environmental conditions are D-brain heavy because a heavy D-brain allocation produces more survival success in a harsh environment. Think Apache Indian or Bedouin tribe member. Conversely a C-brain heavy human will more often develop in a resource rich environment.

Grandchildren and grandparents tend to be alike in C- to D-brain allotment. This allotment zigzagging of human evolution is a way of constantly testing for the correct brain mix to produce maximum survival results in the existing environment. Scientists are missing the importance of this early brain division. There is no free lunch however. The late-stage monkey species and early human species developed to maturity slower and were therefore less capable in early life. They originally possessed only a *small* portion of brain walled off from the D-brain--where it could learn independent of the inherited responses and behavioral patterns.

Lies

Creating this "intra-brain wall" was the evolutionary equivalent of allowing the brain to lie to itself, the origin of denial and lying in worded speech. That was an absolutely necessary adaptation to the first human users of worded speech. In other words, once speech got sophisticated enough to transmit ideas as basic as intention—it was useless if the speaker could not disguise his intent, his motivation, and simultaneously convince the listener that he was being sincere. In harsh times, being totally honest about where you planned to hunt, or how many deer you saw in the field, was probably a starvation or death sentence, depending upon the person to whom you were communicating that data.

Development of this new area of the brain also meant much slower human brain maturity. The D-brain develops itself almost fully before age 5, *and then* it turns over control of the conscious mind to the development of the C-brain. The earlier the D-brain is complete and the C-brain begins developing, the earlier your memories of events—and the more C-brain intensive you are likely to become. This development adaptation more than doubled the time needed for human brain maturity. It forever changed us from our primate ancestors in one fell swoop anthropologically speaking, not a long tedious change like currently theorized.

Slower brain development and additional mental flexibility is a double-edged sword for most creatures. Being "smarter" is potentially a big handicap, since it encourages experimentation with behavior that contradicts inherited memory (reflex behavior coded into our evolutionary rulebook).

Contradicting inherited behavior, for most any reason, is usually detrimental to survival. This is why fully adapted creatures like sharks and alligators didn't need much mental acuity to become so highly successful as species. Their adaptation process, in relation to their environment, is nearly complete. The internal rulebook of their D-brain, which is largely all they have, operates by the distillation of evolution's lessons, not that specific creature's thinking about what to do. Early humans were pathetic predators by comparison to the shark or alligator. They were not near the top in terms of strength, size, or speed. While we developed flexibility to learn information you cannot inherit, it was a costly adaptation. It was so risky that we were the first species (from among a group of "very near" humans) to survive the attempt at using it.

Images to symbols, symbols to ideas, ideas to new images: this is not just a spiritual process, but also a brain process that leads to sign communication and then speech. These were necessary early adaptations that enabled tool making many years later.

Conscious Thought and the Ability to Fool Logic

Though I focus on monkeys, (our evolutionary ancestor), as the example of advanced nurturing, the first adaptive "advanced nurturing" took place in a creature much further back on the evolutionary ladder. This earlier animal created the evolutionary tree limb of animals that care intensively for their young. Monkeys simply took this ability to a higher level much later during the environmentally troubled times aforementioned.

If we travel much further back down the evolutionary tree before intensive concern for offspring emerged, we can find the time where conscious thought, as we know it, emerged. Consciousness is related to sight and images. Human brains "convert" everything our retinas detect (retinas are extensions of brain cells) into "packages" of smaller data that represent the image. This requires changing the collection of light that represents an object visually: into an *idea*, deeper in the brain.

Ideas are much smaller collections of data than the original image, a masterful encryption. The image of two different cups is just the data of the image, twice. But if we reduce the cup to an idea, a thing that holds water, now the brain can recognize a cup it has never seen, knows what it does, and more importantly, eliminates all other image data associated with that cup's image.

Conscious thought is the result of encrypting our visual data into smaller packages of data, built around an idea, not an image. This makes quantifiable objects out of what we see, objects we can then imagine being used for something different.

The development of this encryption process served an important purpose in the vision/self defense center, and simultaneously created what we would consider the first conscious thoughts. Even though this happened hundreds of millions of year ago, the way the human brain works today is just as dependent on this mechanism.

The encryption process is also capable of being fooled. The mind uses the ideas it formed from images to direct its thinking. People have known for years that you can use visual imagery to trick the mind into believing something the person logically knows is false. Some magic tricks and all optical illusions use this process for their visual effect. Recently, this powerful pathway has been beneficially used in medical therapy. People that have constant pain from long-ago amputated limbs can get relief by using a mirror.

A reverse image of their opposing limb fools their brain into thinking the amputated limb is healthy. This story, *The Itch,* is in Appendix III.

Encryption simultaneously allowed the brains of primitive creatures to create ideas, nouns, symbols, and representations. In this way, the brain can focus only on the aspects of its field of vision that are important, that change, and/or are moving.

This encryption process is why some cognitive impairment can impact your ability to find the correct noun in a sentence. Noun words are different from other words by their nature. A noun is a person, place, thing, or idea. In your brain, the words and symbols for person, place, and thing are grouped together, but the idea part (understanding of their use or importance) is distinct and separate. In your brain, the words are titles, hanging on the ideas. When the brain can't find the word representing the idea, the idea is still there!

For example, while concentrating intently during the performance of surgery (an intensely C-Brain task), a surgeon will know exactly what tool he needs to assist him, but may not easily be able to ask for it by name, often choosing to reach onto the OR technician's table and take it.

The brain will see a charging bear in great detail while seeing *nothing else* in the field of vision, unless trained by experts or experience not to react this way. This is why camouflage works until a *part* of the camouflaged object moves; then, the brain associates an idea with the image and sees the rest of the object "appear" out of the background scenery.

Brains must calculate in advance how fast objects are moving and what direction they are traveling before we can physically respond to those objects. This visual brain's anticipation of an object's forward path is what makes surviving possible for most creatures. This cannot be done in a flat visual field.

The early brain had difficulty calculating the speed at which objects in the field of vision were traveling without encryption. We calculate in the first millisecond that a flying object appears in the field of vision, how fast that object is traveling in order to calculate where it will be milliseconds later. This requires more knowledge about the item than just how far it has traveled in the visual field in a set amount of time.

To calculate its continued path accurately, the brain needs to **add** previously learned data to the information streaming in from the eye. Only then can it *finish* its calculation about the object. The idea of a car gives our brain data it needs to calculate its speed, as does the idea of a feather.

There is even a flinch reflex that the brain triggers when the object is approaching so fast the brain cannot quickly label or identify it with an encryption or idea and therefore automatically takes a defensive position. The perception of such a visual threat is followed by reflexive reaction or flinch. Then comes the final D-brain reaction in this chain: the skin or face blush that follows the flinch if the threat doesn't harm us. The last D-brain reaction in this sequence is the D-brain's shame over showing fear. It knows that to physically display fear is potentially dangerous.

For all of above reasons, the brain needed to reduce the data stream. The smaller the data stream, the quicker it can be interpreted in order to generate a successful response. I say interpretation, but that really means calculation. Furthermore, the position/flight characteristics of moving objects can't be accurately predicted (mentally) without this process. In other words, the brain had to develop mathematical abilities in order to see. The right combination of those calculating abilities is the root of consciousness. This is why the fox hides from you, and a fly just flies away when you approach. The fly's eye can see you approach, so it reflexively reacts to that movement, protecting itself by moving. The fox has converted your image and other human images into an *idea of you* and his D-brain knows from evolutionary experience to stay hidden from you even if he has never seen you before--unless his C-brain has learned that you bring food to him daily. The fox in this example has conscious thought -- the fly does not. This is consistent with the Revelation's theory and is consistent with our observation of the world.

It is a balancing act between getting enough information to make a decision and so much information that it slows down response time. In other words, the goal is to gather enough meaning and detail for hunting/work/self protection, but not so much that we can't catch a baseball. We can understand and sense what the speed limitations are in the real world. Basketballs thrown by humans are catch-able. Bullets shot from a gun are so fast that we don't see them or react.

The feedback loop in which the brain recovers ideas from memories in order to compare them to ideas representing the visual images we are currently collecting from the eye was the beginning of conscious thought. The primitive brain, then, learned to advance or "pre-cognate" the scene, in order to anticipate what self-movement would protect the eye (organism) from an inbound missile. That led to consciousness. We still do it exactly this way today.

The foundations of this mechanism in our brain development make it possible for the deep brain to control physical behavior without signaling the conscious mind for control. The deep brain receives its own signal concerning what the eye sees before the conscious mind can identify what it is seeing. It then makes the decision whether or not to duck. This and other clever "tricks" are just two of the tools the D-brain uses to stay "out of sight" and yet powerful, while surrendering enough control for the C-brain to develop. It is the D-brain that ensures we're not sitting ducks for sucker punches, rocks, or fast-moving objects aimed at our head.

One of the central themes of this book is the concept that a decision is made in the base areas of the brain and then makes its way to the "outer" brain, which can only approve, reject, or alter that first decision. The Revelation theory postulates that this interplay is constant and continuous. It represents a brain process designed for and capable of allowing the brain multiple inputs via a second variable: a second opinion separate from the first impulse to act that was supplied by an ancient predator/killer D-brain. This second brain, the C-brain, is born blank, so it can *learn* information, like reading and writing, impossible to transmit through the genetic process. For this theory to be correct, some physical evidence should exist. I was doubtful of the scientific community's ability to demonstrate this brain layering using today's technology. Surprisingly, I found factual support in the scientific community for this theory in a book called The Mind & The Brain by Jeffrey M. Schwartz, M.D. and Sharon Begley. This book is a beautiful treatment of many of the issues raised in this book. I recommend it highly.

Dr. Schwartz introduces us in his book to Ben Libet; a scientist who was inspired in 1999 to follow up on some research done in 1964 showing that the patterns of electrical activity in the cerebral cortex shift just before a human consciously initiates a movement. What Libet then attempted to do in 40 different trials, which has subsequently been repeated and confirmed by other researchers, *is determine when it is that people make a **decision** to act relative to when action is initiated as seen by neuron derived electrical readiness potential.* I will quote Schwartz for the conclusion:

> "The readiness potential again appeared roughly 550 milliseconds before the muscle moved. Awareness of the decision to act occurred about 100 to 200 milliseconds before the muscle moved. Simple subtraction gives a fascinating result: the slowly building readiness potential appears some 350 milliseconds before the subject becomes consciously aware of his decision to move. This observation, which held for all of the 5 subjects in each of the six sessions of forty trials, made it seem for all the world as if the initial cerebral activity (the readiness

potential) associated with a willed act was unconscious. The readiness potential precedes a voluntary act by 550 milliseconds. Consciousness of the intention to move appears some 100 to 200 milliseconds before the muscle is activated—and about 350 milliseconds after the onset of the readiness potential."

For example: when we choose to move an arm, our consciousness of the intention to move occurs 350 thousandths of a second **after** the neurons necessary for the voluntary movement are fired! Naturally this stunned Schwartz, Libet, and many other scientists. They did not consider that the C-brain was a late outcropping of our original D-brain, and thus relegated to editor duties, not initiator. They are still scratching their heads trying to figure out what this all means. Hopefully, the readers of this book won't find this to be a quite as much a mystery. Brains are proof that here on Earth thought can become and/or change matter. Dr. Schwartz clearly recognized the metaphysical connection here between the brain and a higher consciousness, if I read his book correctly.

This C/D setup is why we didn't lose the evolutionarily-inherited information that got us here. That side of the brain, the Darwinian brain side, just partially walled itself off, so it can't be overtly seen by our conscious mind. Still, it did not bury the inherited memory and programming very deep. You may ask, "Why bury inherited reactions and behaviors at all?" We already understand that the D-brain is partially hidden for the purposes of socialization. But, it also was necessary to allow for the establishment of unified command. I'll deal with unified command first.

Imagine a few hundred thousand years ago standing in tall grass somewhere in India, and out of the corner of your eye you catch a glimpse of orange and white. TIGER. In an instant you are secreting adrenaline, your instincts kick in, and you jump, run, fight, whatever. You don't debate what to do with the cooler, more logical, more rationale side of your personality that learned hand signals from Mom in the cave. That side does not contribute to the reaction-- without training. Elite military trainers understand this and train accordingly to compensate for it. The military already takes advantage of the C and D mental division described in this book.

A Military Understanding

The military seems to be the only large organization in the world that applies knowledge about how the C & D layers of the brain interact. They apply these lessons all the way from boot camp to psychological-operations training. The most effective war fighter begins his evolution into a soldier/killer by having his conscious brain broken down, his decision-making thoughts, suggestions and objections blown out and replaced with trained responses. D-brain reactions to obey orders and kill are among those urges sharpened to serve the unit.

This could not be done if these instincts and behavior patterns weren't already present in the D-brain, waiting to be activated. Otherwise boot camp would need to be 13 years long, not 13 (Marines) weeks. The military encourages this brain transition with extreme physical exertion and sleep deprivation to accelerate brain molding. They apply it to all phases of war planning, and always strive for a unified command principle: man-by-man, ship-by-ship, unit-by-unit, and army-by-army. Better a unified crew working under a mediocre commander than two sharp lieutenants fighting for control, with two camps of men backing their favorite. This would be mutiny--and sailors don't voluntarily accelerate at an enemy destroyer while experiencing a mutiny.

One needs to be singular of mind, purposeful, quick to act and scared as hell when face-to-face with a tiger (or a Tiger tank). The DNA instructions of any ancient extinct creature that had the two sides of his brain fighting each other for control at this critical moment became the tiger's dinner. Becoming someone's meal makes reproduction difficult.

Unified brain command was established because those human species that didn't use it were eliminated.

In countries where wild tigers still share territory with humans in the field, the workers wear masks on the back of their heads with big eyes painted on them because they have learned from experience this prevents tiger attacks. This works to protect the field hands because if the wild tiger thinks the human *can see him*, he does not attack. If he thinks the human is turned away, the tiger *will* attack. The tiger does not fear a single human that cannot see him. He has no doubt he can kill and eat a single human. But, if the human can see him--hence the eyes on the mask—the tiger's response is different. The tiger knows that if the human can see the tiger and isn't running, maybe the human is not alone--communicating with other humans or even hunting the tiger. Better to leave him alone.

All of man's physical strengths mean nothing to our long-standing natural enemies. It is our ability to reason, to speak, to communicate, and especially to collaborate, that differentiates us as a species and elevates us to the level of super predator. Predators understand where the line is. The very moment the tables turn and they believe they might become prey, they flee or seek easier food. By their actions, we know our longstanding enemies fear our communication and cooperative abilities more than our weapons.

Socialization

Super predators in general don't socialize well. Try to imagine several lion prides trying to live together. Fur would be flying everyday. The dominant males that keep the pride alive in small groups where dominance roles are clear would cause constant warfare and destruction if multiple dominant males were forced to live together. Their instincts would need to be severely blunted in order for them to live in larger groups. We can easily see that there is too much D-brain and not enough C-brain in a lion. Monkeys have more C-brain than a lion but less than a human.

Advanced nurturing didn't just encourage new types of learned abilities in humans; it allowed advanced adult socialization. Its evolutionary pre-cursor behavior didn't allow elaborate socialization and still doesn't.
The extent of human C- and D-brain adaptation allows a level of lifelong socialization unheard of before. It is this level of socialization that is necessary for advanced tool making. Tool making was the precursor to the invention of the printing press. The printing press has allowed wide distribution of learned knowledge. It was the liberator of the C-brain in humans.

Large-brained herbivores, such as elephants, are not faced with a socialization difficulty, since their food sources don't include meat. Since early humans couldn't exclude dietary meat because of the calorie needs of survival, the human brain *itself* had to figure a way to keep its killer programming while putting it out of sight and out of mind in social situations. Hence, the creation of a "social brain" to communicate and learn things it can't pass through gene inheritance, such as language and science. Images to symbols, symbols to ideas, and ideas to new images: the *Peace Machine* in action.

War

Unfortunately, this hasn't worked as well as it sounds on paper. In fact, man has been fighting himself constantly as long as we have records; in single combat, multiple combat, and organized combat. But without the emerging C-brain, it would have been far worse. If you track what percentage of man's total resources were devoted to war over the centuries, you will see that it *used to be* far worse. I estimate man devoted 50% of total resources toward war in virtually every version of social order tried during most of the time we have been in existence. This holds true from tribal state to nation state--right up until the invention of the atom bomb. After the atomic bomb was dropped, the percentage spent on warfare and defense dropped to around 15% of man's total resources. Now, we are at an all time low of less than 8% of global GDP being expended on war and defense. This is very significant.

The extra capital that was no longer being spent on warfare became the development money behind the agricultural revolution, the transportation revolution, the medical revolution, the computer revolution and the information revolution.

> An alien (non-human intelligence) economist or historian with no prior pre-conceived ideas blinding him, would correctly conclude that atom bombs (used and unused) saved more human lives in the 20th century than did antibiotics, vaccinations and hospitals *combined*.

I predict a future for this point of view in the next history revolution, as historians re-assess man's role as a species in a more objective manner. We have an unlikely mathematical probability of survival unless an unlikely behavioral change overcomes a huge percentage of the world's population soon. We aren't beat as a race yet, but we need to quit staggering down the road, bloody from a drunken fight, while crops rot in the fields, the oil is recklessly consumed, and children are still being starved to death by the millions every year.

This seeming idolization of nuclear weapons and attack on mankind's enlightened nature may appear gloomy, but we can't get to the helpful parts of the Revelation without being honest about the facts first.

> I urge you the reader not to accept anything in this book at face value. Check out everything you find interesting in here. Search for the truth. Question what you are told. With the amount of lying that has been going on for centuries, nothing should be sacred to the truth seeker.

The end of WWII was the beginning of the most innovative, productive, and wealthy period of the human race by the largest margin ever recorded. During this time, the only thing that prevented the continuation of the global war man had been waging on himself in every corner of every continent for as long as we have had records was the *new* threat of nuclear annihilation or Mutual Assured Destruction (MAD).

Many other species communicate well—just not *as* well. This alone differentiates us, and is a by-product of advanced nurturing. Communication abilities are the physical manifestation of the long-term evolutionary direction provided by advanced nurturing (and a ***loving relationship***) between parent and child, usually mother and child. Simple evolutionary pressures and the resultant successful outcomes are self-sustaining. Each incremental increase in communication ability yielded exponential survival advantages because communication/socialization made it possible to *share* ideas, including planning and tool-making knowledge. And, good hunting plans and tools are the keys to a successful hunt.

Book I, Chapter 3

A New Brain Inside The Old One

Lions and Tigers and Bombs

As a result of the transfer of command to an alternative thinking center (C-brain) that didn't operate with pre-set reactions during non-emergency times (more flexible behavior), human survival improved—but only a little. Primitive men, the one using pre-set reactions hard wired into the D-brain by evolution or the one beginning to experiment with learned information in the C-brain, had only limited success against his competition. Humans didn't dominate the Earth until just a few thousand years ago, a geologic blink of an eye.

> This newness of our ascension as the dominant species is important to understand, because most of our modern thinking is controlled by buried Darwinian programs which *were shaped* by millions of consistent hits by the evolutionary hammering tool which I have labeled "Brutal/subsistence experience." Only a few thousand hits were by new environmental influences I have labeled "Indulgent experience."

Because of the recent ascendancy of C-brain dominance in conscious thought, we are far more controlled by the brutal subsistence programming than by social experience, material indulgence and education. We now live in a dangerous time, when the influence of the Stone Age D-brain is still very strong but our hands hold weapons fashioned by our C-brain toolmaker. "Tools" powerful enough to kill us all.

This recent combination of advanced nurturing and the walling off of the inherited memory from conscious thought turned out to have greater adaptation benefits than any other evolutionary alternatives explored by nature to date. This process was very risky to the species at first. Before writing was invented and complex tools existed, this adaptation may have been tried many times before by other branches of our species...and failed. It may still prove to be riskier to human survival in the long run than if we had stagnated without the adaptation.

After all, how could individuals learn enough from their fellow primitive creatures to enhance their survival over that of animals who used only the inherited programming perfected by eons of evolution? Here's where our weaknesses finally paid off. Man's relative weaknesses in speed, strength, biting power, hearing, and especially in smell forced our ancestors to rely more on their developing mental abilities. Without special physical skills or camouflage, man had to rely on his brain.

Thinking and communicating at a more advanced level within the human species was probably only a bit better for survival than the skills of other similar species. It allowed him to form alliances with animal species that needed our help but had superior skills in a particular area (dogs/smelling, horses/portage, etc.).

These small bands of our modern ancestors experimented with crude tools (spear and arrowhead) and a communication-based hunting model. This emerging 2-brain adaptation fought for existence side by side with the "lions, tigers and bears" for millions of years through Ice Ages, droughts, pandemics, and volcanoes. Humans and their domesticated partner species were not extinct but were doing little more than just surviving.

We were mostly hungry during this period as well. This is the reason the human body does so well with so little. Humans are designed to weather lean times. If the average adult male eats only 1700 calories a day, significantly below the government recommended levels, it will increase your life expectancy by 10 to 15 years and extend the number of good, mobile, relatively pain free years by 20.

Because of the low number of years humans have lived with plenty, as opposed to the millions of years we lived close to starvation, we never developed an evolutionary protection mechanism against overindulgence. Compare this to black bears, whose gluttonous behavior before hibernation does not threaten its organ system the way too much food threatens ours. Widespread obesity is the expected result for us. Humans in advanced countries are the first group to have hit the food trifecta: low cost, abundance, and an amazing variety of foodstuffs. They are also the first group of humans to eat so much food containing processed and unnatural food ingredients, that they are poisoning their own bodies. Our brain evolved during near continuous starvation experiences. That evolutionary pressure resulted in our extremely efficient fat and calorie preserving human machine. The result in the rich countries is an obese population that continues to eat and eat—driven by a D-brain impulse which wants to store fat for a rainy day—no matter how much our C-brains don't want our bodies to do this.

If you live in the U.S., I am sure you remember standing in a grocery store while your C- and D-brain argued over whether you should buy a bag of potato chips or a chocolate bar. Your C-brain knew rationally that you didn't need the calories, but you were urged to buy it anyway. It's the D-brain that supplied the urge to take out your wallet and buy the junk food. It's not just the flavor. The D-brain craves food it *knows* are full of calories, a hedge against future starvation. Food companies understand that the average consumer is either D-dominant and wants calories/sugar/carbohydrates or is C-dominant and wants NO calories, sugar, or carbohydrates. Most advertising is designed for these two groups. It is modeled according to how our brains *respond*. Ironically, neither choice alone is healthy.

Our recent victory over the other super-predators was a squeaker. Mitochondrial DNA analysis proves we all came from a single common female ancestor, or Eve, as the anthropologists like to call her, who lived just 160,000 years ago. This suggests numerous close calls with extinction. There were many other species of "near human" type creatures whose entire line became extinct. The surviving line of humans we belong to was likely a greater factor in the extinction of these other lines of "near humans" than were any environmental factors. Even in Eve's descendants, it's likely that some people--women, children, sick, injured or old--were mostly prey for the vast majority of this time. We only shot up from the back of the evolutionary pack to lead the super-predator race recently.

Think of the Earth during early human society as a pyramid with only the top 20-25% of the species in the predator pool and the other 75% of the species reliant upon them (the top 25%) for food and protection.

Like this:

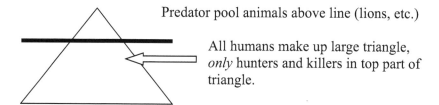

Predator pool animals above line (lions, etc.)

All humans make up large triangle, *only* hunters and killers in top part of triangle.

Prey pool animals (like deer) outside the triangle and human tribe members (inside the triangle) make up the part below the line.

Up until approximately twenty thousand years ago, the human population was essentially a small static number. These pre-writing, pre-Copper Age humans had only skills, which allowed them to survive, not dominate. The split in brain functionality took a long time to progress to the point where the smaller blank slate conscious mind was sophisticated enough to perform with consistent superiority above our combined enemies.

The D-brain matures first in humans for a reason. If the environment is harsh enough, there is no need to waste more than the allotted minimum brain capacity and maturation time on the C-brain. It is the D-brain that has the best chance of keeping the organism alive in a tough environment.

For 99.9% of the time man has been on the planet, lions and tigers have been our mortal enemy. *Now, we visit them at the zoo.* Ironically, animals at the zoo maul people each year because we have let the C and D barrier get so thick that some people fundamentally fail to understand the threat these animals represent. They stick their hands, legs, feet, and cameras into the enclosures.

Think for a second how scared you would be if you were locked in the cage with those same tigers. That's who you really are. Imagine them leaping over and eating you alive; that's who they really are. So, who are *we*?

Male versus Female Brain Split

At the moment of sex selection in the fetus, there is a further brain dichotomy pushing our split brains into two further divisions. This sex hormone-induced change in brain wiring creates two types of humans that behave differently and have different capabilities.

Instead of dismissing this idea, ask yourself one question. Do the men and women you know appear to act differently and have different capabilities if you were to break them into the two camps by sex?

> Sex differences are not about intelligence, which is fairly even. Science has now shown that levels of sex hormones in the human fetus alter the development of neural wiring that persists into the adult brain.

Men and women are like two equal computers. Each sex is *loaded* at conception with both operating systems; but at some point the human computer must choose to run one operating system or the other. Looking at sex differences, do you think they are conditioned into humans by society or that sex behavior differences occur naturally? Thinking about the children you know will make it very simple. If you take away a little boy's toy guns, he will just use sticks as toy guns. If you give him dolls instead of guns or sticks, he'll just turn them into make-believe weapons or combatants. Little girls do not usually behave this way. They dress little girl dolls up, and *nurture* them.

Little boys are conditioned by most of our developed societies NOT to be aggressive -- or to use violence to maintain order in their world, *but they do it all day anyway*. As little boys, their D-brains are rehearsing for the big day, the real hunt.

When adult humans fail socially (criminals), it is often because they fell back on these traits (killing, lying, stealing, physical abuse/torture) when stressed because that's how their D-brain programming tells them to behave. A small number of people by random variation are D-brain killers by birth. Often, they are further conditioned to it by circumstances. They exist, living among the rest of us, hiding their secret urges/actions until they are caught.

As a species, we have millions of these super-killer wolves living among six billion killer sheep. Still, more of those sheep are starving to death than are being killed by the "wolves." The starvation events are caused by natural and economic conditions that are exacerbated by genocidal human leaders who desire the death of minority populations living under their control, and often use the natural events as cover for their killing campaigns.

Homicidal humans aren't just world leaders. They live in our neighborhoods as well. I knew a man that tried to hire a hit man to kill his wife. I was in Rotary with him. Never really liked him, but I was still surprised. I've shaken hands with lots of people that have killed other humans, and a few that have killed or ordered the killing of thousands.

Criminal Activity and Murder

Actually, the urge to kill a spouse or ex-spouse is understandable—from a D-brain point of view. In a case like this, the *Peace Machine*, or C-brain will actually help the D-brain (e.g. Scott Peterson) kill a spouse.

The D-brain male/female bond includes sex, which is very bonding. Betrayal here is maddening, and many times bloody as a result. The patterned behavior in this example is so predictable that it actually can lead you to the criminals! Most crime scenes reveal a pattern. Most patterns tell you very specific information about the perpetrator and victim. This quirk is why the behavioral sciences profiling unit of the FBI at Quantico Virginia is so successful. They understand the relationship that D- and C-brain programming have with murder and emotionally-motivated crime.

They can successfully create "profiles" because most crime scenes reveal enough of a pattern to paint a picture of the interplay between the D- and the C-brain of the assailant at the time of the crime. This often gives away his or her relationship to the victim.

People kill people they hate, have had sexual relations with, and/or to whom they are emotionally bonded. Through the use of the D-brain, they stab them more times than necessary, for example. Or, the murderer will cover the dead body with a sheet, literally the C-brain hiding the D-brain crime from the C-brain's view. Conversely, people have the consent of the D-brain when they kill with their C-brain. But C-brain murder is usually for money from someone you don't know or don't care about emotionally. The C-brain plans more than the D-brain.

Enough of a brain pattern can usually be observed at a crime scene to deduce race, sex, age, strength, rage level, and proximity to the victim socially. Even more can be figured out in the hands of an experienced profiler. Criminals can't help it. The D-brain in most criminal situations carries out the killing. It may use C-brain plans, but since the D-brain is in charge at the killing moment, the crime scene is like a roadmap to the killer's D-brain motives, rage level, and *opinion* of the victim.

Interestingly, the FBI stopped at profiling and only applied what they were learning about D- and C-brain behavior to crime, leaving the rest of the non-criminal population unstudied from this perspective.

In my opinion, the top profiler alive is in Florida. Dayle Hinman. She is a genius. When she walks into a room and says, "Oh no, this wasn't a professional hit, this was done by an angry woman," or "This was done by a man with a criminal past that has had contact with this person," she is almost *never* wrong. Watch some of her TV shows, **but not all of them.** You'll be mesmerized. Read this book, twice. Watch some more episodes you purposely didn't watch the first time and, and *you'll* be guessing the identities of the criminal--before the ending is revealed.

D-brain Urges

If a human can engage in a conscious activity that is approved by society (hunting for example) that is also a D-brain favorite, the person will generally get great satisfaction from that activity. News commentators were openly confused when Michael Vick was arrested for torturing and killing dogs used in dog fights--admitting on air their intellectual conflict about why they felt Michael Vick was a bad man for participating in dog fighting, but felt no rage toward their own relatives *who hunt and shoot deer regularly.*
Few of these commentators mentioned the bigger question: why did Vick do it

at all? This is important because the government evidence showed the dog fights only had a few thousand dollars *in total* wagered on them. Buying dogs, feeding dogs, and maintaining a secret place to do this cost more than could have been won; and, Michael Vick certainly did not need the money. *He did this because he enjoyed watching them fight—the D-brain loves blood sport.*

There is also a genetically programmed racial aspect because dogs were not used as hunting companions in Africa as they were in European countries. The D-brain is OK with killing prey that represents food such as deer. In fact, it likes and rewards this behavior with positive emotional satisfaction. Ask a hunter who stocks his freezer with venison.

Dogs, on the other hand, are a domesticated species that have been living as partners with humans (in some areas of the world) long enough (100,000 years) that our D-brain considers it a crime against our own tribe to kill one of these human co-partners without good cause.

In many warm climate areas (like Korea), where dogs have *not* been man's hunting companions, they are on the menu at restaurants and are served in homes. No problem here with the D-brain.

Humans will kill all other species, indiscriminately, almost without thought.

I'm sure many Americans stood around their kitchens talking about what a bad man Mr. Vick was for killing/fighting those dogs while standing just feet from a refrigerator filled with red meat oozing the blood of another animal, be it pig, chicken, or cow. Those same humans sat down chewing the meat on their dinner plate *as* they expressed rage at Mr. Vick.

I'll bet the judge who sentenced Mr. Vick eats meat. He's probably even hunted and shot animals he subsequently *didn't eat.* This is not hypocrisy on the judge's part. He feels no guilt over the sentence he imposed. Guilt is the rejection one side of the brain feels when the other side negatively contemplates what the other side said and/or did.

The judge's D-brain considered dogs a companion species worthy of protection, and his C-brain approves of laws protecting that relationship. So he experienced no conflict, no guilt, and no hypocrisy. While his brain still likes to kill animals, his ancestors had a history different from that of Vick's ancestors, so the groups of animals it is OK to kill, in his mind, doesn't include *dogs.* The *Peace Machine* was hard at work in Mr. Vick's legal case:

lots of denial, racism, and base behavioral programming, all swimming around together in our head with our C-brain beliefs--that we aren't killers.

Men/Women...The Difference

People think because men and women *should* be equal in the eyes of the law that they are *actually* equal in all respects. The word "equal" is a bad term from a biological and evolutionary point of view, and is the main reason so many people get this wrong.

If you needed two different tools to get a job done, any description with *equal* in it would automatically be wrong. Survival of the species was dependent upon the critical increase in productivity that results from the use of two different tools--a division of labor and responsibilities within the family structure. Hundreds of other species use this sex-based division of labor because it gives them **increased survival odds.**

Any process that undermined the male/female division of labor and priorities (*equally* important) made humans less healthy as a species and was de-selected by evolution. Men and women react differently, think differently, care about different things, and behave differently *in every culture on this planet.*

This is primarily because women deliver and nurture babies. Men either support this role, or do not. But in no culture of humans on Earth are these rules/roles reversed.

Although women inherit all the intellectual capacity and D-brain programming that those men do, their hormones and eventual sexual maturity activate different sets of behavioral neural programming.

This results in significant behavioral differences that enhance *our survival as a species.*

Both sexes get all the inherited behavioral programming; they just don't use all of it. Homosexuals and trans-genders get one brain hormonal activation code and a different set of physical parts. Both sexes fear death and both have a strategy for attempting to defeat it. To deny this is unfortunate.

D-brain programming instructs the female to count on

> her mate for food, security, sex, and shelter. The programming wants her to have a baby(s) and raise it. D-Brain programming instructs the male to get *satisfaction* (a gentle emotional high) from hunting/working, activities which provide him with meat/money. His programming instructs/expects him to trade this meat/money for sex, food prep, outfitting, child rearing, and asset protection. His brain does not instruct him to rear the babies, although he can activate these programs if forced or encouraged, just as a woman can activate the hunting/working routines if forced or encouraged.

According to statistics worldwide, a woman raises at least nine out of 10 human children. Once a man must raise children, the instructions can become activated. But, hormonally speaking, the programming activated by child rearing is heavily female. This may sound obvious or boring, but what follows is not widely known.

Love and Manipulation

The female C-brain makes another decision during early puberty that splits the girls from that point forward into two more sub-groups, based on the level of nurturing they received.

> Girls raised a certain way (highly nurtured) have a higher probability (compared to males and non-nurtured females) of choosing love as *the* tool to orchestrate the necessary trades for subsistence. Only a minority of females and a smaller minority of males -- pick this path. All the other girls choose verbal, social, and physical (sex) manipulation as the tool set to orchestrate the necessary trades for enhanced survival during their lifetime.

The idea that these behavioral patterns are learned from the environment is completely false. The way men and women manipulate each other is remarkably consistent race to race, and nationality to nationality.

The reason is that humans are all so biologically similar and heavily influenced by our inherited programming; it couldn't be any other way. Only the outward expression of a person's internal programming, looks different culture to culture.

Manipulation, for the purposes of discussion in this book, is not intended to have a negative connotation. This warning is so important, it will be repeated in the next chapter.

Males generally don't make the Love vs. Manipulation choice at this young age because the development pattern they inherited doesn't force them to choose this early. They can continue to rely on the D-brain a few years longer. Males relied upon might for so long as a species that male brains do not understand until they are older that they also need manipulation skills. Scientists, educators, and parents interpret this as boys "maturing" slower than girls.

Primitive men controlled their environment by force. Manipulation was a fallback primarily used on *other men*--humans they couldn't necessarily dominate with violence.

Primitive women, because of strength and behavioral objective disparities, started using manipulation much earlier in life. Hence, they advanced the skills of manipulation further in their hormone-induced brain programming. The C-brain, the *Peace Machine,* in both of these sexes is still just a tool of the D-brain.

Most of the time, the D-brain gets its way.
Remember this rule about the D-brain. It is an important one in the *Peace Machine* theory and in your life. Understand your D-brain motivations. Understand your C-brain actions. Accept that people around you will mostly behave according to their D-brain objectives, no matter what their C-brain controlled mouths are saying.

If you visit the local courthouse for a few weeks, you'll see a lot of evidence to support the above point as regards the "good" and "bad" outcomes of D-brain activity. People get married at the courthouse mostly because they are in love, want to make babies, and want to assure their mate that they're not even thinking about other mates. Of course, it's also where the murder trials occur for spouses who later decide to kill each other. Understand that your D-brain uses the information it gathers when you are a child--to make the big choice about what set of behavioral programs to use throughout the rest of your life...the "Love" set or the "Manipulation" set. The brain will then pick one of these two and start to use the associated programming to select everything in life from whom you will date (and/or have sex with) to picking a career and choosing between jobs. These choices are highly stressful activities for most humans because it irritates the C- & D-brain relationship. The C-brain believes it should be making the choice about what job to take. After all, it's the toolmaker, and has been bringing home the bacon the last few thousand years.

The D-brain can't read or write, but it understands that job choice involves money (deer meat), which heavily influences success at survival and reproduction, *its primary jobs*. So, people struggle to make choices as they weigh the desire to do one job for emotional reasons versus salary and perks. Since these battles happen simultaneously in both the C- & D-brain, people can get very confused when it comes to career and job choices. Unless they're desperate--then the D-brain shuts down all opposition and makes a choice centered on survival/security.

The D-brain ultimately makes the decisions in the areas it thinks are important, regardless of the C-brain. This is why physical, sexual, and pheromone attraction or "chemistry" between two otherwise incompatible C-brains usually wins out. Many people like this marry each other – leading to the saying "opposites attract." This is one example of a D-brain decision for which there currently is no other rational explanation.

In second marriages, where people vow to find a mate more like them in temperament (C-brain compatible), they often slip up and the D-brain gets them involved with another C-brain incompatible spouse because of D-brain attraction. Modern psychology hypothesizes this is due to one person trying to fix another person, because the first person had a loved one with a similar flaw. That explanation is complete nonsense. Observe the humans around you for a few weeks using the C- & D-brain model to interpret their behaviors instead of the current model you use. It will be like watching the world through x-ray goggles.

Back to the "Love" set of programs versus the "Manipulation" set. Essentially people who pick the "Love" set attempt to structure their social order to receive and give love and deer meat with all the humans they desire to be around, depend upon, or who depend upon them with no specific quid pro quo.

These D-brain "Lovers" are primarily female. They believe in love at the very core of their brains because they accepted nurturing and/or aspire to it. They believe that basing their behavior on love is in the long-term interest of themselves, their children, and other family members. This causes them to act a certain way. They provide love generously--and material support as available--without an emotional or material price, unless they genuinely deem "pricing" the support as *constructive for the other person.*

Humans who pick the "Manipulation" set (the majority), use all the same cultural and physical norms, like marriage and sex, but do so operating from a different perspective. They know about love; they just don't trust it. They observed the humans around them growing up. They carefully observe humans who base their core decisions on the "Love" set and take note of the fact that even "Love" people have C-brain manipulation machines on their shoulders.

So for a collection of reasons, the "Manipulation" D-brains finally assume everyone else is, at the core, willing to manipulate anything to survive. This leaves their D-brain no choice. It won't pick a pattern it thinks is inferior to another pattern for survival.

Naturally, in homes where there is a lot of dysfunction, manipulation is chosen more often than not. Even in homes lacking major dysfunction, where multiple children are born, the brains will generally alternate (not every other one necessarily) enough to maintain variety.

The eldest is most likely to pick love, and is the most likely to succeed in life financially and at having more children. The second and remaining brains born into a family *are aware from birth they have sibling competition for resources,* an influencing factor in which programming set is chosen.

The early C-brains who pick the manipulation set of programs then begin to manipulate every sentence, every utterance, every effort, counting on and attempting to instigate in the humans around them reciprocity of some sort, ranging from emotions such as love, praise, flattery, respect, and attention, as well as actual material reciprocity. A lot of what goes on during puberty is the flexing of this programming—for the first time.

Ironically, every C-brain thinks it is actually making choices based upon something altogether different. It believes its actions are based upon something you learned in your C-brain *long* after birth: right and wrong, moral and unethical, legal and illegal, good and bad, honest or lying, logical or stupid, with God or against God.

None of these C-brain validations of our actions make much difference in the end, because the C-brain value system is just a front for the agenda of the D-brain. When people surprise you with conduct that appears to make no sense, examine it from this perspective—and it will clearly make sense.

It's the D-brain *desires* that consume 95% of your time, money, energy, and attention. Manipulator-based D-brain goals and love-based D-brain goals are very similar – it is only the behaviors used to attain these goals that differ.

The primary goals every brain tries to accomplish are to survive and reproduce. These goals persist whether it picks the manipulator set or love set of patterns to use.

Take a few minutes to review memories and experiences of your life and you will see many examples of the truth in this statement in your life to date. What they have stumbled upon below is the outline of the Peace Machine *and* its limitations.

> Undecideds More Decided Than They Think, Study Says
>
> By BENEDICT CAREY
> Voters who insist that they are undecided about a contentious issue are sometimes fooling themselves, having already made a choice at a subconscious level, a new study suggests.
>
> Scientists have long known that subtle biases can skew evaluations of an issue or candidate in ways people are not aware of. But the new study, appearing Thursday in the journal Science, suggests that professed neutrality leaves people more vulnerable to their inherent biases than choosing sides early.
>
> Experts say the findings might help explain why political polls can be so off-base, and why some people make up their minds in the voting booth with little sense of why they pulled the lever yea or nay, blue or red.
>
> In the study, researchers asked 132 residents of Vicenza, Italy, where they stood on a ballot issue being debated at the end of last year: the expansion of a local United States military base. The residents also performed a computer test in which they quickly clicked one key for negative words flashing by on the screen and another for positive ones, and sometimes words were paired with images of the base.
>
> People unconsciously in favor of the base tended to take longer clicking negative word associations, and vice versa.

By tracking the number of errors and the time it took to answer, the test estimated a person's implicit positive or negative associations with the base.

Asked about their stand a week later, 9 of the 33 residents who had declared themselves undecided were now in favor and 10 were opposed. The scores of those 33 people on the computer test the previous week predicted which way they would turn, said one of the study's authors, Bertram Gawronski of the University of Western Ontario in Canada.

Experts said the study might not mean much for the coming United States presidential election, in which a whole list of issues is in play. But it does put exit polls in a different context.

"Voters explain their reasons by relying on cultural and idiosyncratic causal theories that may bear little relation to the real reason for their preferences," Timothy D. Wilson and Yoav Bar-Anan of the University of Virginia wrote in an editorial that accompanies the study.

End of article.

"Voters explain their reasons by relying on cultural and idiosyncratic causal theories that may bear little relation to the real reason for their preferences."

I couldn't have said that better myself. For the most part, humans, regardless of their actions or beliefs, or in whose name they're acting--God, country, or passion--are actually just carrying out behavioral programming developed and inherited during a million years of evolutionary survival lessons from their ancestral parents.

In conclusion: Humans act out their D-brain programming fear of death, desire for money, sex, nesting, babies, etc--using C-brain rationalizations that they believe are the real reasons for their actions and desires.

Humans acting on D-brain programming believe their rationalizations are the real reason for their actions—because of Denial.

Males and Love Motivation

The male brain can choose love as a primary motivator. But this usually happens later in his life, after his D-brain love patterns of behavior are awakened by pre-set environmental triggers such as: a loving relationship (unconditional love) from a parent or important adult, a successful sexual/mate/love relationship and/or most importantly---the birth of a child. Most fathers will admit that the birth of their children changed them more than any other event(s) in their life.

These pre-programmed male D-brain triggers are aimed at converting our hunter (the C-brain-equipped one focused on weapons and hunting), into a provider for the new family. These triggers need to be pushed at the last second, because loving, considerate men *without families* would not have survived to reproduce as well in the brutal/subsistence environment *nor would they have been as attractive to females in that environment.*

Males don't "mature slowly." They develop advanced nurturing characteristics when the babies arrive—any earlier extends the length of time the males are in a weakened state defensively. Women must develop advanced nurturing first because they make the nest. Female nesting behaviors are a manifestation of inherited memory pushing the C-brain out of the way. The D-brain is highly concerned with this event and cannot afford to leave it to chance learning.

Men inherited this nesting knowledge and behavior as well! The brains of some men—whose wives are pregnant--become "convinced" they are pregnant too. They then undergo a wide range of *physical*, emotional, and behavioral changes associated with being pregnant.

Please re-read this entire chapter carefully (men and women) if you plan to have a successful marriage during your lifetime. Especially focus on what is to follow. Most of what you need to know is contained in it. *It may sound crazy, sexist, and politically incorrect, but it's true.*

The brain patterns activated in females by their hormones are designed to facilitate, motivate, induce, seduce, and lock in a mate to provide for her and her children. Some feminists will feel about this section the way that the Pope felt about Galileo. However, truth is our goal. This male/female split has worked for a long time and generally lines up both sides of the brain toward the same goals as shown in the table opposite.

Notice in the tables below that C-brain goals have changed in very recent times, causing potential internal conflicts in the C- and D-brain relationship that are leading to increased marital conflict and divorce.

The Revelation

ANCIENT MALE
D-brain Goals   C-brain Goals
1. Stay Alive   1. Hunt / Deer Meat
2. Have Sex   2. Trade Meat for Sex

ANCIENT FEMALE
D-brain Goals   C-brain
1. Stay Alive   1. Nest
2. Have Babies   2. Mate

Ancient males and females alike used denial and manipulation as C-brain tools to achieve the above. However, the male D-brain used killing to achieve its goals and the female D-brain used nurturing. This is the biggest difference between the ancient sexes. In the next chart, please note what has changed for the modern female.

MODERN MALE
D-brain Goals   C-brain Goals
1. Stay Alive   1. Make $$
2. Have Sex   2. Trade $ for Sex

MODERN FEMALE
D-brains goals   C-brain
1. Stay Alive   1. Make $$
2. Have Babies   2. Mate/nest

Modern males & females still use denial and manipulation as C-brain tools to achieve the above. However, the modern male D-brain doesn't have to kill to achieve its goals, and the female C-brain is now also expected to "hunt for deer meat" or earn money.

These changes appear subtle but are not. Our societal expectations as decided by our C-brain have changed while our D-brain programming remained the same. As we see above, staying alive is the big idea. It's accomplished through fear of death on one hand, versus sex (and everything connected to it) counter-balancing that fear. The male D-brain uses the C-brain to moderate its behavior and to use "advanced hunting techniques" to get along with female brains. This explains the "metro-sexual," an otherwise strongly heterosexual male with a highly-developed "female heavy patterning" of his C-brain.

This "female" brain pattern is more helpful to a modern man who doesn't need to kill to eat anymore but increasingly does need to interact with greater and greater numbers of female C-brains. Conversely, women are migrating toward the male pattern since they can become more aggressive, without having to physically back it up in modern society. Unfortunately, this creates competing C-brain goals among the sexes, and, worse yet, competing C & D brain goals in modern females.

Sex *used* to be contradictory. Before birth control, it decreased your individual chances of survival by placing huge demands on your resources and diverting some of the hunter killer instinct to nurturing. But collectively we don't survive without a next generation--the outcome of sex. Consequently, we are programmed to do it even though it contradicts the first urge of staying alive because of faster resource usage.

Remember the duality principle described above. As we will see, duality applies everywhere else as well: two big forces pulling in different directions, both necessary for existence. That's the true nature of the universe.

The duality principle describes not only our brain function but is systemic throughout the Universe.

This is further explained in later chapters. There are always just two powerful forces pushing back and forth at the root of everything. This is true: in the Sun, economics, thermal dynamics, politics, philosophy, your brain, your life, sex...you name it.

Female/Male Conflicts in Modern Times

A very important and obvious point in the charts (opposite page) that bears repeating: modern technologically advanced societies created a split between the motivations of the modern C-brain female and her own D-brain. This creates significant internal tension. Furthermore, females are now competing with males' C-brains to make money. This creates external (interpersonal relationship) tension. These two things are a double tragedy for woman.

Internal C- and D-brain conflict can wreak havoc on the emotional health of any afflicted human. Indeed, many more women in industrialized societies are suffering emotional illness than are men. The Muslims are acutely aware of this--and it serves as proof to them that God is on their side. These points of tension also create an imbalance in the system, which we see today as disintegration of the family unit. **Systemic societal problems are always the result of the C- and D-brain primary forces shifting out of balance.**

The external tension resulting from competitive male/female C-brain goals, combined with the female dysfunction caused by internal tension, has created waves of turbulence in the male/female relationship dynamic expressed as high divorce rates in technologically-advanced societies.

When We Changed the Speed of Evolution

> Whether you are an Einstein or a stay at home parent, our lives are basically a tug of war between two forces beneath the surface. We have many more tools today than did primitive man. Still, an infant from Christ's or the Pharaoh's time could come here, be educated, and no one would be the wiser.

It was over millions of years that evolution weaned out inferior traits, literally killing off the carriers of bad behavior programming before they could reproduce. It also rewarded the "good" thoughts and traits that displayed evolution's successful principles through extended survival of their offspring.

These slow lessons are burned into our genetic record. But, change on this pace is glacial. Writing and reading, *coupled* with printing (distributing the effect), changed the pace of our C-brain evolution dramatically. This was the beginning of the new age. Now symbols and images are being displayed and understood by billions of humans.

> *The past few hundred years* are the first moments in Earth's history that a complicated artificial device, the printing press, affected the evolution of a natural creature's brain. This was the moment the slow moving substance that was humanity could start multiplying *effort times productivity*, increasing worldwide GDP. Economics and science blend with the millions of C-brains waking up to their productive power as they learn to read. This change altered the balance of power between our two brains, in billions of brains simultaneously. This is the biggest change in human history.

Writing didn't *just* lock in the knowledge base; it actually forced the brain of the author and reader to push the size, importance and development of the C-brain. Evolutionary selection also supported this progress in incremental steps. The brain didn't have to get bigger by cell count. For the C-brain to expand, it can simply expand into areas currently used by the D-brain. This re-allocation of assets is made possible due to the excess capacity the D-brain developed over the many millennia of human development. A computer analogy will help explain this. Suppose a programmer used a particular hard drive and memory to do many different tasks. The computer would build up excess but unused instructions/data, in those areas. We de-fragment our computer hard drives for this reason. The human D-brain also has the ability to "de-fragment" during development, and can allocate that excess capacity to the C-brain.

The developing human brain needs a lot of environmental stimulation before the brain decides how to shape itself. It goes through this process twice, once for the D-brain and once for the C-brain, usually over 21 to 25 years. The D-brain didn't get rid of all its primitive knowledge; it just stored it more efficiently in case it's needed again, not completely out of reach. Our brain dynamically combines the storage of information *with* processing ability--something a modern computer cannot do, or you would never have to re-boot to change the kernel.

The D-brain won't completely hide its information to the point where it couldn't be used; otherwise, the creatures experimenting with the freedom and flexibility that C-brain dominated thoughts allow wouldn't have survived even a hundred years. So, the D-brain uses adrenaline, among other "tricks," to actually take over the mind and body in an emergency. The D-brain also uses emotional triggers/motivations (pride, ambition, love, hate, greed, etc…) to point/drive the C-brain mind toward decisions it favors during non-emergency times.

Many people testify in court about having no memory of committing certain crimes, usually ones that are violent and/or emotionally driven. Nobody ever believes them. I do. I used to watch trials as part of my duties as a reporter. After seeing things repeated so many times, I realized that such a frequently recurring lie, told in such similar ways, was not statistically probable. Interestingly, many people terrorized by crime also have limited memory of the attack, but everybody *believes* them because the current dogma is victims "suppress" these memories rather than storing them in a different brain area.

Suppression of this vital information is not logical. Why? The brain would want data collected during trauma to be the most accessible—and it is—to the D-brain. Many assault and rape victims, who have no conscious memory of their attacker at first, later recognize them. The D-brain can recognize just a snippet of voice, even a smell.

These people aren't lying in the traditional sense when they claim initially to have no memory of the attack--It's a *Peace Machine* lie. Many people don't know about the things they do when their adrenaline is pumping. The part of the mind/brain we return to after the adrenaline abates--is not the part that was in charge at the time of the crime/outburst/unfortunate remark/or attack. It is the side of the brain that is *in charge* at the exact moment of an event that has the initial memory.

If the C-brain was in charge during some event, the memory is "converted" to long-term memory and stored in the D-brain *and* in the C-brain memory banks once you go to sleep. If it's unimportant, it is discarded, and neither brain remembers it consciously. This conversion process is an electrical-to-chemical conversion. Unfortunately, this path doesn't always run both ways. An event that happens when the D-brain is temporarily in charge is not necessarily stored in the C-brain. Furthermore, the D-brain discards a lot of what the C-brain thinks is important. Rote memorization is C-brain repetition designed to alert the D-brain to data importance, so it will not discard it.

You have probably said things in your life that if confronted with later--you would deny saying. But "go the tape" as they say in Vegas, and you would see something surprising: you would see yourself making remarks your C-brain would deny.

You can use information about this phenomenon in your own life. The next time someone tells you what you said—in anger or a heightened emotional state—don't argue that you never said such a thing. Write down what they report you said. Sit down and read it a few times while trying to mentally place yourself where you were physically at the time of high emotion. See if your mind/brain can find the statement you believe you never made—coming out of your mouth. Almost **every** person is guilty of this *and* blind to it. If you think you are an exception to this--you are in denial.

This occurs because the conscious mind, which resides in the C-brain side of the brain, wants to believe it's in charge. Ironically, this false pride is a trait it is goaded into by the D-brain, the home of pride. So, the C-brain creates a "false" history, a fairy tale that explains how and why the conscious mind is in charge.

To do this, the C-brain first denies the other side of the brain exists. Additionally, the D part of your brain, the one that makes all the important decisions in your life, doesn't care that you are reading this book. It wouldn't even care if it understood you might learn how and why it does what it does—because it knows it's in charge and always will be.

"There shall be no other God before me." The first commandment, universally it seems. I'm not making light of this principle. **But, it's very important to know that all our laws and societal patterning are the result of our brain structure—not the other way around.**

Almost every individual you know believes they are consciously responsible for all the actions and decisions they make in life—job, behavior, voting, marrying, and where and when to buy a home. Actually, it is the D-brain attempting to fulfill programming installed by millions of years of relentless evolutionary honing that is making every one of those decisions with two primary goals that push and pull against each other.

1. Don't die/get killed. *Money or deer meat represents life, hence its importance.*
2. Reproduce or sex, sex, sex. *Self-explanatory.*

The primitive mind doesn't understand sperm or condoms. Sex is so important to men and women because our D-brains want children; and, the act of sex is how our brains think we get them. As a result, sex strongly bonds people chemically (Oxytocin) to the person with whom they are engaging in sex, or else you would kill them when food got scarce, or when enraging triggers like adultery or insubordination surfaced, and the gene line would die. Actually, humans do kill people close to them for these very reasons. C-brain restraint or logic isn't always strong enough in time to stop D-brain behavior motivated by fear of death/sexual jealousy, rage, and/or desire.

This is extremely important information concerning your life. Living with, having children with, and having sex with, another human cements the bonds we call love when properly understood and nurtured. However, when not properly understood and nurtured, these same actions will cause divorce, heartache, murder and mayhem. We like to think we fall in love for the stupid reasons we read about in magazines and on the Internet. We expect these "reasons" will "cool" after marriage. This is untrue because the reasons themselves were rationalizations created by the C-brain. **Of course they "cool," they weren't even there in the first place.**

Adultery is rooted in brain architecture. Males look to have sex with *other* women after establishing a nest with a primary woman, because the programming in their D-brain wants them to increase replication of their DNA. Remember, the D-brain almost always gets what it wants. Married men fool around more than married women because they don't fear getting caught as much as women do. This is a hangover in attitude from the "Might is right" environment, where adultery didn't get men killed as often as it got females killed. Interestingly, modern environments are changing this dynamic for men and women, as both sexes move toward the middle in the industrialized areas of the world where female brains are expected to hunt deer meat and male brains are forbidden from controlling their environment with violence, thus forcing them to manipulate more. We have seen, and will see, a higher and higher female adultery rate. The male D-brain is aggressive (in its pursuit of sex partners) because it cannot read or write. It doesn't understand the concept of sperm and thinks we are impregnating the female(s) *with the sex act itself.*

This is why men's libido is so high when they are successful and so low when they are under financial stress. Their D-brain knows it can support more children than the one-nest/one women can provide for practical purposes, while times are good. Men with lots of power and money almost always have multiple sex partners, even multiple families, as a result of this behavioral set-up. This is just simple biology. We have a tremendous amount of evidence supporting the primary importance of this concept in evolution in our environment; we simply use denial to block most of this data.

Love, not lust, almost *never* comes first. Any adult who is honest will admit to an experience of real love for the first time through the birth and nurturing of a child. Matrimonial love is built into the man/woman relationship *as it grows* by previously identified evolutionary triggers like childbirth, not a perfect night on the town in Paris. The following sentence is one of the most important things you will read in this book. A healthy family life is the *only* path to long-term brain happiness. Understanding what creates dating/mating emotions in the brain can spare you much heartache, and help you create a successful family.

The family you form is the only source of lasting satisfaction and love.

The brain is structured this way regardless of what your parents, books, or anyone tells you. Only your brain can make you happy. These are the triggers that please the brain. Drugs, including alcohol, are an exception, but their euphoric effect is temporary. They are the exception to most everything in this book because they directly affect the brain cells.

Drugs are a kind of short circuit. They glide past the brain-blood barrier and disable the brain's customary defenses. A rat in a cage with access to a lever that distributes euphoric drugs or alcohol will push the lever continuously, without eating, in the case of some drugs, until he dies. People aren't quite as bad about this as rats, but we're dangerously close. See chapter 9 for more on drugs and alcohol and your predicted/expected struggle with them.

The Nature of Attraction

Initially human-to-human attraction is a lure set by nature to get us into a family situation. The mistaken assumption as to the importance of romantic love, a mistake perpetuated and propagated in Western industrialized society, has literally ripped apart the social fabric in the West, leaving everybody with more questions than answers. Lust and love are not the same thing and aren't located in the same part of the brain. However, one *can* lead us to the other, and the product of both is excellent for the continuation of the species. The dual role of lust in the D-brain and love in the C-brain is good for human family creation and advanced nurturing. The Pope missed this point is his encyclical "God is Love," blaming 'Eros" or erotic love (sexual love) for man's division from God. This is not accurate, and encourages some women (hardly ever any men) to think sex is a sin and should be avoided even in marriage (except for procreation).

Women and men who withhold sex or act against the core brain programs, usually as a manipulation tool, but sometimes because of drugs and alcohol, invariably end up destroying their relationships. Many unhappy D-brains use these same techniques to force the end of a relationship—often without C-brain awareness.

Meanwhile, many overly-aggressive people, mostly men, leave their gene line (generation after generation) for others to rear because they are too aggressive to stay and do the job themselves – or, they don't survive to do it. The super-aggressor's value to the tribe in a hunting and gathering format was much higher than it is today. Today, for example, football players still get enormous benefit from this programming setup, but exhibit many of the conflicts that it causes with modern society. Another example of super-aggressors today is the warrior class, a subset of the population pool, which is still essential for a sharp military. They are a necessity for our national species to fight wars and defend the race/religion/country. A country can actually get this group killed with aggressive national behavior or a technologically lopsided conflict, and may lose its sovereignty as a result. It is the coupling of the denial mechanism with C-brain flexibility and D-brain drive that fosters a tool-making, ambitious, highly flexible species. Our species then used those tools to push food production to the point where a few hundred thousand subsistence hunters and gatherers around the world blossomed into a population of over 6 billion humans in less than .0001 percent of the time they've been in existence as a species. And, because of our denial mechanism, we remained ignorant of the process that got us here or the understanding that we have two disparate yet conjoined brains making decisions for us.

In the run-up to modern times, the dual purposes of the male and female brains created social unification though these purposes were outwardly dissimilar. Humans think of themselves as individuals. But in reality, there are only two core types of adult humans: humans in a partnership with mate(s), and humans not in such a partnership. The latter state is actually physically harmful to the brain as an organ and can dramatically affect behavior.

The combined "brain" of the mated couple will operate differently, both for the man and the woman, compared to an un-partnered brain. As this book describes in detail, the two sides of the male or female brain operate to please the other mated brain to the degree to which it is practical to do so. The man who sheepishly ducks out on his drinking buddies halfway through a night of carousing, to their open taunting, after his wife calls him on the phone--is likely in a healthy relationship. The evolutionary purpose of commitment, co-dependency, jealousy, behavioral programming, desire, instinct, and emotion is to produce offspring—whether or not the C-brains of the humans in question realizes it.

> We are much less honest than we need to be as a species concerning how important it is to the emotional health and development of a human to have and raise children.

We have the luxury of ignorance on this because our birth control technology has far outpaced our evolutionary brain development. It should be noted that raising children without having birthed them, imparts the same protections and emotional maturation that being a natural parent does, provided the person undertaking this role wants to do it. This "Adopter" behavioral pattern can be especially kind, loving, and generous. I suspect this is a D-brain compensation mechanism for physical, perceived or voluntary non-fertility. Conversely, the unfortunate "Cinderella experience" is also possible when the person adopting the child is doing so non-voluntarily.

Marriage rates across all races in the U.S. have being falling for decades. I consider the U.S. and certain European populations a leading indicator for other world populations because of their economic success. One group of people is still getting married regularly however—couples hoping to have kids. The need to be mated creates a brain dependency-deficiency in non-mated humans, except in non-mated humans that are **not** potentially active sexually because of impairment or age.

> Ideally, you are ½ of a complete man-woman, or "mated," brain unit. While single (not in a committed relationship), you aren't equal in brain capability to people who are mated. Please don't get upset by this and take it personally. Think of it as a math problem. If two variables are more productive because they job share, like a wrench and a screwdriver, then it is a mathematical certainty that one of them will do one job well, and will *not be* well suited to the job for which the other tool is designed. Separate the tools and this automatically dictates that *both tools* will now have things they cannot accomplish at maximum efficiency.

Evolution is about efficiency. It's not sexist rhetoric. It is evolutionary reality. Recognizing this gives you power over it. Many of the Revelation's concepts contradict public policy, theology, philosophy, and popular opinion. This is a completely different way of looking at yourself and humankind.

No woman or man is complete as a family, as an evolutionary unit, or as a complete brain unit without taking this into account. Very little mention is made of healthy brain co-dependency either scientifically or philosophically. But it is clear that men and women operate better, more productively, more lovingly, and for a longer time, both in total years and healthy years, if they live in this mated/married model. Ask an insurance agent. The cultures that completely embrace these ancient brain roles generally remain unified socially: Muslim, Tribal societies, fundamentalist, polygamist, etc. It must be pointed out that these same cultures appear quite backward socially to the C-brains of individuals in the rest of the world. Our modern C-brains may not understand why women can't drive in Saudi Arabia. However, the vast majority of Saudi women are not complaining.

Book 1, Chapter 4

The C & D Brain In Your Marriage, In Your World, And In Your Universe

The C- and D-brain structure has many modern day effects, not all positive. This brain split in areas of social responsibility is one of the underpinnings for the war between the U.S. and radical Muslims. This is true despite the religious, social, and political C-brain rationalizations spouted by both sides. The D-brain urge, and consequently the C-brain plan, of the male power structure in extreme fundamentalist Islamic societies is to kill any people who have evolved past that controlling philosophy. The D-brain is making the decisions in these communities and nations, and has a written C-brain rationalization (Islamic or Sharia law) for killing anyone espousing a philosophy that is a challenge to the Islamic religious principles. There is little or no conflict for true believers. The D-brain urge or purpose is hidden behind the religious rationalizations.

Our C- and D-brains in the U.S. *were not even aware* of the conflict between the way extreme fundamentalist Muslim brains see Western society and the way our Westernized brains view Islamic societies. If you don't believe that statement, ask yourself a few questions about the morning of September 11, 2001. Did your C-brain know *before* the second plane hit the second tower-- but after the first plane hit--that we were under attack?

After the second plane hit, did you have *any* idea who had attacked us? After the government said it was al-Qaeda, did you have any idea who that was or *why* they were mad at the U.S.? Did the spell checker dictionary in your computer at that time even have the word al-Qaeda in it? Two of the largest newspapers in the U.S. still disagree on how to *spell* al-Qaeda.

Western-thinking countries (C-brain dominant) with their democratic ideas about power-sharing with the voting public, a man discussing his Christian faith with his own son, and educated C-brain active women in Saudi Arabia that want to drive automobiles are all examples of a challenge to the D-brain male power system.

In most fundamentalist Islamic societies you can still get the death sentence for doing or promoting any one of those three "crimes." No other written explanation, except the one in this book, describes why this is logical. Most non-fundamentalist, non-Muslim people think: "They're crazy" -- as if that alone is a worthy explanation.

When women in Saudi Arabia want to drive and have a say in their country's laws or leaders, will the U.S. support them or the Sheik who is married to multiple women and controls the oil? One more layer of the onion, one more war, one more D-brain and C-brain conflict. Remember the power of denial and the fact that this is a two-way street. Many foreign leaders routinely engage in practices, like bigamy, that are punishable by jail in the U.S.

U.S. Presidents routinely meet and negotiate with leaders from oil rich countries that violate our C-brain ethical rules. But these same Presidents will not meet with world leaders from countries lacking oil reserves that embrace customs and practices the President considers morally repugnant. "Human Rights Violations" is the main C-brain rationalization used by U.S. leaders in this situation. Meanwhile, *in the* U.S., the FBI and other governmental officials will jail you and take away your family for the exact same practice of bigamy that is practiced by the world leaders we embrace and literally hold hands with.

The U.S. government only respects, talks about, and highlights cultural differences when it benefits the government's agenda, or when those differences can serve as a useful political wedge. If more important interests like oil, food, trade, money or even military cooperation are at stake, the C-brain sensibilities get to take a back seat to D-brain priorities. The very old question: "Dear God—Why?" concerning the U.S. government's support of tin-horn dictators the world over, including *Saddam Hussein*, is answered quite nicely here. Hypocrisy, self-interest, denial, and C- and D-brain lying, are all rolled together here. This is not a judgment, even though words like these ordinarily carry a heavy negative emotional connotation.

To attain a true understanding of your world and truth you need to move past those old ways of looking at things. Many Americans are aware of this seeming hypocrisy on the part of their government and are angry about it. A popular book during the Vietnam War era, *The Ugly American*, characterized this as American hypocrisy. The actual source of the problem is rooted in our evolutionary past, and its resultant effects on our laws, traditions and power structure.

The U.S. didn't lose the Vietnam War because we are hypocrites. If that were the case, nobody would ever win a war, since humans are all hypocrites. The problem is that the C- & D-brain split can divide the will of the people. It can especially separate the D-brain-heavy people from the C-brain-heavy. That's how you lose support for a governmental policy or action. If the U.S. had been as committed to defeating the Vietnamese as they had been to defeating the Japanese, Hanoi would have fallen in the first year of the war. There is a big reason the U.S. has won only one large war, Gulf War I, decisively since WWII. The C- & D-brains were united in that war's objective. As such, it was accomplished quickly and with relatively low loss of American life and treasure.

The D-brain dominated person is more likely to be Republican. At this time, it is the War and Security Party. The C-brain dominated person is more likely to be a Democrat. It's currently the Peace and Fairness Party.

In General:

D-brain Republicans support: defense spending, individuality, self-reliance, tax decreases, less government, fewer laws and regulations, hunting, gun rights, opposition to marriage between anyone except a man and a women, opposition to abortion, and an anti-immigration movement.

C-brain Democrats support: lower defense spending, alternative dispute resolution like the UN, restrictions on individuality through higher taxes and social programs and regulations, and more public administration of critical infrastructure like health benefits. This agenda is a perfect match for C-brain priorities, our *Peace Machine*.

The C-brain invented every tool we have that has made survival easier for humans in the last 5000 years; and it knows how clever it is scientifically. This makes the C-brain falsely confident in its ability to control its social and economic environment through legal manipulation. C-brain dominance, does correlate at the voter booth with higher education levels, as would be expected by Revelation logic.

Despite the commonplace expectancy that such a correlation would exist there is **little** religious voter correlation because it doesn't correlate at the C- & D-brain level. Both brains believe in God, the D more than the C. Our C-brains can doubt the existence of God, but never our D-brains. An excellent analysis of the 2004 Presidential election demonstrates this, and in fact, ties together a number of C- & D-brain characteristics empirically.

My thanks to Jonathon Schechter, who put the following study of the 2004 Presidential election on line at the following URL for Revelation readers:

http://charture.org/manager/uploads/2004%20Presidential%20Voting%20Analysis.pdf

I intended to insert some of the data from this excellent study here, but I couldn't do it without taking up too many pages. I urge you to go to the URL provided by Mr. Schechter and study these charts while using the book as the study guide. As you study these charts on the Internet, ask yourself one critical question: do the voter patterns more closely correlate to the current political dogma or to the C & D patterns described throughout this book?

This should be on the cover page of Mr. Schechter's research when you arrive if you've found the right spot:

U.S. 2004 Presidential Election Voting Patterns
Religious Activity and Education Levels as Correlates of Voting Patterns
November 8, 2004

The split you will see in the charts displayed on the Internet represent the C- & D-brain split, as it existed at the time of the 2004 election in the U.S. You can clearly see how far apart the two sides had drifted. You would have to go back to WWII, to find a time when the collective consciousness of the nation was truly unified in its C-brain and D-brain thinking.

A country's public loses wars when they aren't brain unified, not because of particular military outcomes, cultural hypocrisy or technologic superiority. Most people (and the Pentagon) would argue that whether you lose or win particular battles determines the outcome of war. I don't agree.

> A large, rich, and powerful country can lose a war just because the domestic population at the D-brain level no longer considers the enemy a real threat. If, concurrently, the C-brain decides the loss of life and treasure (deer meat) to continue the fight is an unacceptably high price, support for the war effort erodes, and ultimately the effort is lost.

The U.S. lost more battles during WWII than in Korea and Vietnam combined but won that war--and did not win the latter two. Some people might nitpick here, and say we didn't lose Korea, but I propose that Kim Jong-Il is proof the U.S. didn't *win*. For this reason the U.S. government actively waged a domestic propaganda campaign during WWII designed to keep hatred for the

Germans and Japanese at a fever pitch.

The C Brain Chooses Manipulation Techniques

The societal and genetic information lays our first foundation layer: the C- & D-brain goals. The next thing to be decided by the brain during its development is which technique to use to obtain the goals. This decision is *equally* important. Let's begin at the individual level.

> Our C-brain, the *Peace Machine*, always chooses manipulation as its tool of choice in both males and females. There is no other option. It was created by the D-brain in the first place to do this. The C-brain is the social instrument of the entire brain, created by the D-brain to use denial and manipulation as well as advanced communication, tools, and temperance to **aid** socialization. This is the *Peace Machine exposed*.

Remember all those fighting prides of lions? We know we don't act like that anymore, so there must be a reason we don't. It's an adaptation humans have that lions don't. If a human can think about the idea of *not killing* the other animals like itself but instead cooperates with them, that human must be thinking with something else...something new, an adaptation new to meat predators. Our herbivore beginning, before the diet change, contributed to our C-brain beginning. The social programming allowed by herbivores grew with the extra "leisure" time gained when monkeys added the high calorie supply from meat eating. What resulted was a new way to get sex, other than violence; a way to get food, other than violence; and a way to get along other than through violence or intimidation.

The C-brain was the D-brain's finest achievement. It's the thing with which we contemplate enlightenment. Of course, even while it's doing that...the D-brain still wants to have sex. It used to just take sex, and is still urged at some core level to do this. In some men, when the C-brain temperance control fails or is destroyed, they revert to un-tempered D-brain programming and rape for this reason. Read 50 case files on how serial rapists behave and you'll recognize a pattern by the tenth file. These men go on the hunt, target a victim, stalk the victim, and then move in on the woman as if she were prey. Doctors and mental health professionals claim rape is violence-driven, and that is *true* because the D-brain is also the violence center. If you use the D-brain to get sex, it will probably be violent sex. The C-brain, on the other hand, learns the sweet nothings, which must be murmured to obtain sex in a more socially acceptable, C-Brain controlled manner. Physical force/intimidation was a much bigger D-brain set of tools in ancient times; hence, the primal male reliance on rape, physical intimidation, and violence inside the male/female relationship.

> The big question for each specific human: did the D-brain, as a child, choose the "Love" set or the "Manipulation" set of behavioral programs as the basis for accomplishing its goals through the C-brain? This early choice, of which humans are not conscious, has a big impact over the course of each human life as you accumulate the fruit each of these brain patterns bears. Neither choice is right or wrong.

When you currently make decisions, the C- & D-conflict is represented by a "Does the ends justify the means" kind of internal debate in your mind, or maybe a "To forgive or not to forgive" kind of choice if you are religious. Remember, these conscious C-brain calculations are just covering for a decision your D-brain wants. **Manipulation, for the purposes of discussion in this book, is not intended to have a negative connotation.** Please re-read that sentence a few times. Many people stumble here. Manipulation can be taking a beating from the man *so the children don't*.

…It can be self-righteous anger not genuinely felt or perhaps genuinely felt, but expressed a certain way to get a desired response from another human(s).

…It can be the trade of sex or emotional comfort for food, money, or shelter.

…It can be a facial expression designed to elicit a certain response.

…It can be table image for a poker player.

> Manipulation, then, is the vast number of psychological, physical, emotional, and mental actions, thoughts, words, and deeds that humans perform to manipulate other humans into doing what they want the other person to do.

This is highly productive behavior. Denial and lying are essential to the success of this behavior. In order for this behavioral pattern to be successful—

The Revelation

and nonviolent—it simply requires more lying and more manipulation. That's why humans are so capable of these behaviors. As an experiment tonight, I want you to try to think of just one situation that currently exists in your personal life that might be different from *The Peace Machine* view. To get you started, I will provide an example:

Susie wants Fred (husband) to make more money. They fight about this a lot. He likes his job teaching English. She wants him in computers, where her best friend Jane's husband is working--*he's the one making the big bucks.* They have been around and around this many times. His reaction is always emotional; hers is always "bitchy" and insistent. His complaints to his friends about her are not flattering. Most importantly, Susie and Fred perceive themselves as not getting anywhere on these issues.

By conventional standards, this couple is headed for divorce court. A *Peace Machine* examination of their problem leaves room for a fix. Susie is *programmed* to push the hunter responsible for deer meat (money), even though her conscious mind is not even aware of why she does this. When she was more or even completely dependent on him for subsistence in ancient times, this was a necessary/good thing for her to do.

Remember: humans lived lives of subsistence for a long time. Conservation of resources and constant attention to getting more were the keys to survival. Nowadays, Fred takes this personally because his Darwinian brain reacts unhappily to her questioning of his:

1. Authority
2. Hunting ability (Earning power)
3. Self esteem

In the old model he could be pushed, but his authority would never have been openly questioned because he ruled by might, an option less available to him today in industrialized nations and communities. In reality, Susie loves Fred. She's *had a baby by him and this has tied her brain to him in ways she will never be able to undo.* She pushes him because she has chosen him as the hunter for her and her young. This type of behavior—while annoying to the male--is actually a compliment. Susie doesn't know this either. This is why she can't express to Fred how she really feels without it "coming out wrong." Fred, on the other hand could handle all these conflicting signals if he knew what motivated them in Susie's D-brain and how to address them while talking to her C-brain. Some complaints people make are valid, but many of them are the result of our genetic programming running its course regardless of need. Why? That is how the human mind works. As a result of our evolutionary progression, the inherited behavior patterning will be obeyed minus any interference from a conscious effort to re-pattern.

The damage to Fred's D-brain confidence by Susie's comments hurts his **swagger** with Susie. Susie, by virtue of Darwinian programming is drawn to strength (might was right for so long) and is actually repulsed by Fred's weakness (although she won't say this nor admit it to herself--denial). Their behavior is a death spiral for their marriage. One of them has inadvertently started a cycle, which the other has inadvertently perpetuated to their doom as a couple. The hope is that some knowledge about the forces behind all these actions and reactions can change the result.

Susie didn't think about all this. She was a cheerleader. The inherited software contained inside human female D-brain worked well for them in the primitive environment that molded our existence for 99% of our time on Earth. At the edge of the ball field in high schools across the country, scantily dressed young women gather to entice the strong, fast, young football players into physical relationships. In furtherance of the D-brain programming, this results in any number of them getting pregnant right after the game despite the availability of modern birth control methods.

Only one male high school football player in every one hundred schools will make it into professional football (the real goal line) and earn enough income (deer meat) to feed a family through that activity. Here the denial mechanism is supporting D-brain priorities and behaviors--bad for *both* sexes. Why do I say it is a "bad" choice? At the same time the cheerleaders prioritize the long shot football players, they completely ignore the library where all the computer programmer boys are gathered--future millionaires, good providers, non-abusive, understanding, very manipulate-able.

We live in a new environment; millions of years of programming can't be re-written by a few hundred or even a few thousand years of different experiences. New environments take time to translate into new programming. Women have been conditioned for millions of years to get sexually and emotionally aroused when they come up against aggressive, strong, arrogant, physically violent men. They are not aroused at the sight of pocket protectors —not yet.

Some women's C-brains catch on after the "Geek's" money (deer meat) arrives. And, these women will get blind to age, physical appearance, and overcome their initial repulsiveness to mate with these men. This is possible because the female brain is pre-programmed to run down the list of potential mates by preference order, and find one with which it can fulfill it's D-brain mission. It can then construct sexual desire for that choice even if that choice is a form of monkey, such as a Bonobo. I know this sounds difficult to believe, but modern research has already bumped into this set of facts. Although this set-up will appear logical to a student of the C- & D-brain arrangement, it contradicts all other modern thought about sexuality so completely that the female study author who discovered this fact admits she simply has no explanation.

I have severely shortened an article about this topic, which appeared in the *New York Times* on January 22, 2009. The entire article is worth a read. I urge you to look it up on the Internet.

What Do Women Want?

By DANIEL BERGNER
Published: January 22, 2009, New York Times

Meredith Chivers is a creator of bonobo pornography. She is a 36-year-old psycholgy professor at Queen's University in the small city of Kingston, Ontario, a highly regarded scientist and a member of the editorial board of the world's leading journal of sexual research, Archives of Sexual Behavior. The bonobo film was part of a series of related experiments she has carried out over the past several years. She found footage of bonobos, a species of ape, as they mated, and then, because the accompanying sounds were dull — "bonobos don't seem to make much noise in sex," she told me, "though the females give a kind of pleasure grin and make chirpy sounds" — she dubbed in some animated chimpanzee hooting and screeching. She showed the short movie to men and women, straight and gay. To the same subjects, she also showed clips of heterosexual sex, male and female homosexual sex, a man masturbating, a woman masturbating, a chiseled man walking naked on a beach and a well-toned woman doing calisthenics in the nude.

While the subjects watched on a computer screen, Chivers, who favors high boots and fashionable rectangular glasses, measured their arousal in two ways, objectively and subjectively. The participants sat in a brown leatherette La-Z-Boy chair in her small lab at the Center for Addiction and Mental Health, a prestigious psychiatric teaching hospital affiliated with the University of Toronto, where Chivers was a postdoctoral fellow and where I first talked with her about her research a few years ago. The genitals of the volunteers were connected to plethysmographs — for the men, an apparatus that fits over the penis and gauges its swelling; for the women, a little plastic probe that sits in the vagina and, by bouncing light off the vaginal walls, measures genital blood flow. An engorgement of blood spurs a lubricating process called vaginal transudation: the seeping of moisture through the walls. The participants were also given a keypad so that they could rate how aroused they felt.

The men, on average, responded genitally in what Chivers terms "category specific" ways. Males who identified themselves as straight swelled while gazing at heterosexual or lesbian sex and while watching the masturbating and exercising women. They were mostly unmoved when the screen displayed only men. Gay males were aroused in the opposite categorical pattern. Any expectation that the animal sex would speak to something primitive within the men seemed to be mistaken; neither straights nor gays were stirred by the bonobos. And for the male participants, the subjective ratings on the keypad matched the readings of the plethysmograph. The men's minds and genitals were in agreement.

All was different with the women. No matter what their self-proclaimed sexual orientation, they showed, on the whole, strong and swift genital arousal when the screen offered men with men, women with women and women with men. They responded objectively much more to the exercising woman than to the strolling man, **and their blood flow rose quickly — and markedly,** though to a lesser degree than during all the human scenes except the footage of the ambling, strapping man — **as they watched the apes.** *(Emphasis added)*. And with the women, especially the straight women, mind and genitals seemed scarcely to belong to the same person. The readings from the plethysmograph and the keypad weren't in much accord. During shots of lesbian coupling, heterosexual women reported less excitement than their vaginas indicated; watching gay men, they reported a great deal less; and viewing heterosexual intercourse, they reported much more. Among the lesbian volunteers, the two readings converged when women appeared on the screen. But when the films featured only men, the lesbians reported less engagement than the plethysmograph recorded. **Whether straight or gay, the women claimed almost no**

arousal whatsoever while staring at the bonobos. *(Emphasis added.)*

"I feel like a pioneer at the edge of a giant forest," Chivers said, describing her ambition to understand the workings of women's arousal and desire. "There's a path leading in, but it isn't much." She sees herself, she explained, as part of an emerging "critical mass" of female sexologists starting to make their way into those woods. These researchers and clinicians are consumed by the sexual problem Sigmund Freud posed to one of his female disciples almost a century ago: "The great question that has never been answered and which I have not yet been able to answer, despite my 30 years of research into the feminine soul, is, What does a woman want?"

Full of scientific exuberance, Chivers has struggled to make sense of her data. She struggled when we first spoke in Toronto, and she struggled, unflagging, as we sat last October in her university office in Kingston, a room she keeps spare to help her mind stay clear to contemplate the intricacies of the erotic.

"No one right now has a unifying theory," said Julia Heiman, current director of the Kinsey Institute for Research in Sex, Gender and Reproduction. It's important to distinguish, Heiman said, between behavior and what underlies it. Heiman questions whether the insights of science, whether they come through high-tech pictures of the hypothalamus, through Internet questionnaires or through intimate interviews, can ever produce an all-encompassing map of terrain as complex as women's desire. But Chivers, with plenty of self-doubting humor, told me that she hopes one day to develop a scientifically supported model to explain female sexual response, though she wrestles, for the moment, with the preliminary bits of perplexing evidence she has collected — with the question, first, of why women are aroused physiologically by such a wider range of stimuli than men.

Are men simply more inhibited, more constrained by the bounds of culture? Chivers has tried to eliminate this explanation by including male-to-female transsexuals as subjects in one of her series of experiments (one that showed only human sex). These trans women, both those who were heterosexual and those who were homosexual, responded genitally and subjectively in categorical ways. They responded like men. This seemed to point to an inborn system of arousal. Yet it wasn't hard to argue that cultural lessons had taken permanent hold within these subjects long before their emergence as females could have altered the culture's influence. "The horrible reality of psychological research," Chivers said, "is that you can't pull apart the cultural from the biological."

Still, she spoke about a recent study by one of her mentors, Michael Bailey, a sexologist at Northwestern University: while fM.R.I. scans were taken of their brains, gay and straight men were shown pornographic pictures featuring men alone, women alone, men having sex with men and women with women. In straights, brain regions associated with inhibition were *not* triggered by images of men; in gays, such regions weren't activated by pictures of women. Inhibition, in Bailey's experiment, didn't appear to be an explanation for men's narrowly focused desires. Early results from a similar Bailey study with female subjects suggest the same absence of suppression. For Chivers, this bolsters the possibility that the distinctions in her data between men and women — including the divergence in women between objective and subjective responses, between body and mind — **arise from innate factors rather than forces of culture.** *(Emphasis added.)*

Ultimately, though, Chivers spoke — always with a scientist's caution, a scientist's uncertainty and acknowledgment of conjecture — about female sexuality as divided between two truly separate, if inscrutably overlapping, systems, the physiological and the subjective. Lust, in this formulation, resides in the subjective, the cognitive; physiological arousal reveals little about desire. Otherwise, she said, half joking, "I would have to believe that women want to have sex with bonobos."

Besides the bonobos, a body of evidence involving rape has influenced her construction of separate systems. She has confronted clinical research reporting not only genital arousal but also the occasional occurrence of orgasm during sexual assault. And she has recalled her own experience as a therapist with victims who recounted these physical responses. She is familiar, as well, with the preliminary results of a laboratory study showing surges of vaginal blood flow as subjects listen to descriptions of rape scenes. So, in an attempt to understand arousal in the context of unwanted sex, Chivers, like a handful of other sexologists, has arrived at an evolutionary hypothesis that stresses the difference between reflexive sexual readiness and desire. Genital lubrication, she writes in her upcoming paper in Archives of Sexual Behavior, is necessary "to reduce discomfort, and the possibility of injury, during vaginal penetration. . . . Ancestral women who did not show an automatic vaginal response to sexual cues may have been more likely to experience injuries during unwanted vaginal penetration that resulted in illness, infertility or even death, and thus would be less likely to have passed on this trait to their offspring."

Evolution's legacy, according to this theory, is that women are prone to lubricate, if only protectively, to hints of sex in their surroundings. Thinking of her own data, Chivers speculated that bonobo coupling, or perhaps simply the sight of a male ape's erection, stimulated this reaction because apes bear a resemblance to humans — she joked about including, for comparison, a movie of mating chickens in a future study

When she peers into the giant forest, Chivers told me, she considers the possibility that along with what she called a "rudderless" system of reflexive physiological arousal, women's system of desire, the cognitive domain of lust, is more receptive than aggressive. "One of the things I think about," she said, "is the dyad formed by men and women. Certainly women are very sexual and have the capacity to be even more sexual than men, but one possibility is that instead of it being a go-out-there-and-get-it kind of sexuality, it's more of a reactive process. If you have this dyad, and one part is pumped full of testosterone, is more interested in risk taking, is probably more aggressive, you've got a very strong motivational force. **It wouldn't make sense to have another similar force. You need something complementary. And I've often thought that there is something really powerful for women's sexuality about being desired. That receptivity element. At some point I'd love to do a study that would look at that."** *(Emphasis added)*

End of NYT/Chivers article.

So, even though women are capable of mating with a monkey (Bonobo) or the computer geek if nothing else is available – their hearts (D-brain) are not in it, not at first, anyway. Some might even marry the computer geek or the rich old geezer, but they'll always love the Neanderthal on the motorcycle who either actually slapped them around or displayed a strong, violent side. Their D-brain programming instinctively glues them to that "bad boy" (or even just his memory) with epoxy. Might was right for a really long time. Evolution will eventually change this. Put another way, women and their offspring survived by staying close to the strong, mating with the strong, and bearing the children of the strong. As the above article demonstrates, short of this questionable objective (mating with violent men) in today's world, they also carry the inherited D-brain capacity to generate sexual readiness with a much broader range of mates than the male brain does.

To further elaborate and illustrate this point, without the monkey factor, I will now quote, from an article that ran in *The Washington Post* on Sunday April 20th, 2008, page M2. This is an actual letter to an advice columnist named Carolyn Hax.

> Dear Carolyn,
> My wife and I have been married four years. We share a mortgage but don't have kids or other significant debt. My wife works a lot harder than I do. Her company pays her $100,000 a year, but she is always exhausted. I have a publishing business that pays me $150,000 annually. I have been building my business since before we married and now enjoy the passive income it provides us.
> My wife is resentful that she has to work so hard and she sees me kicking back. I would love to travel by myself once in a while or do a guy's trip, but I get nothing except guilt from her, which in turn, makes me angry and resentful. It feels like there is a constant cycle of resentment because of it. She stays in her job because there is potential to move up, and because she enjoys the challenge and the responsibility. She is also making terrific contacts—she likes working hard. I've always told her that if she doesn't like her job, I support

> anything she would choose to do, regardless of her income.
>
> I feel I carry my weight financially (and so does she). Shouldn't I enjoy the fruits of my labor without feeling guilty, and shouldn't she give me the freedom to enjoy it once in a while? She has vacation days she can use if she wants. I would need a second job to make more money, which we don't need right now. She implies that I am lazy and not driven. I disagree; I built my business with hard work and drive. Doesn't my income count heavily toward that argument?   J.L.

I want to include Carolyn's response before I give my own because without any knowledge of the *Peace Machine* or this brain division, she sums up her answer with a very penetrating *Machine* thought.

> I suppose, but I would make a different argument entirely: that being driven is seriously overrated. Certainly I'm glad certain people are. We all enjoy—in fact, take for granted—countless fruits of other people's elective 80-hour weeks.
>
> I simply reject the implication that it's necessary, or even desirable, for everyone to be driven. People pulling elective 80-hour weeks certainly enjoy—the fruits of other people's rejection of that life.
>
> It's not just poets, volunteers and people who make sure they have nothing more pressing to do than walk at their toddler's pace. It's people who think 40 hours more than suffice.
>
> You have a pretty sweet life. Whether you earned it or picked it up off the sidewalk is, I think, immaterial. You are content with what you have. If your wife envies your contentment, then she needs to do something to find more – with your cooperation of course. Her insistence that you lessen your contentment, by taking on equal stress, of all things, is appalling. A stunningly selfish solution.
>
> Granted, you don't mention any ways that you apply your spare life quality of life toward improving hers – chores, cooking, social planning, to cite a few examples. If you don't do this, then do this. "I support anything she would choose to do" isn't a promise kept only in few possible futures, it's one to make good on daily.

> If you do already pamper her though, and a thoughtful, happy, well-paid spouse isn't enough to make her happy, then it's time for you both to start asking what is.

This letter, and Ms. Hax's response, sum up modern relationship thinking. This is probably the reason she got the job. Before I dissect the thinking on all sides here, I want to get to Ms. Hax's last sentence:

> "If you do already pamper her though, and a thoughtful, happy, well-paid spouse isn't enough to make her happy, then it's time for you both to start asking what is."

Exactly. The wife here isn't happy with everything the way we say it "should be." Why? This man's wife has a brain. Her D-brain wants a baby. Her C-brain doesn't even know this. Her D-brain also wants a strong, domineering mate that goes out on the hunt everyday–regardless of how much meat is already in the cave. Deer meat didn't come in lifetime annuities or royalty checks in ancient times.

She is imitating this male behavior herself, because he isn't doing it. Her C-brain is confused about how, in today's society, to make her own D-brain happy! Remember, he said she was exhausted by all this work. People who work 80-hour weeks when they don't have a financial imperative do it out of passion. That is not exhausting. This woman's motives for working long hours are purely a D-brain need.

I don't know if Ms. Hax has babies. If she does, she should have included her feelings about this in her answer. If Ms. Hax admitted that having them was the single most important thing she has ever done, the only one that gives her lasting satisfaction and contentment--and that the thought of her man and those babies, make up 90 percent of her total contentment, *her answer would have really helped this poor guy out.* This does not appear to make sense on the surface. However, the vast majority of Ms. Hax's brain, and this unhappy wife's brain, are dedicated to a process she never even mentions in her answer.

This letter and her answer illustrate a C & D point about how unaware we all are of our base patterning and genetic programming. Why does this man's wife differ from Ms. Hax or any other female? She doesn't, of course. No disrespect meant to Ms. Hax, but does she really think it makes no difference to this man's wife whether he *found the money they live on* lying on the sidewalk or earned it? *He* is content. His *wife* is not.

The Revelation

Taken together, these two ideas are actually very funny. It is bizarre to think that whether he is content or not would influence her contentment. Whether he lucked into easy money or was a reliable earner makes a huge difference to a mate. If she has a plan in her brain for a family—which I contend she does—it makes a big difference whether he lucked into a herd of wounded deer and brought home a meat bonanza, or is looked up to by the rest of the tribe as a consistently successful hunter. Her D-brain knows it will take years of reliable meat deliveries to feed her offspring.

As a result, the woman in this letter is not content. It does make a difference to her. If she isn't sure he is a hunter, protector, *killer*, willing to fight every day for her and their (as yet) non-existent kids in the daily struggle for deer meat, her D-brain will be very irritated by his behavior. *This happens whether her C-brain even knows whether or not her D-brain wants to have a baby.*

He is playing the role of the modern, successful entrepreneurial man... confused about women to a T. **And, he will end up without this wife as a result**. My apologies: to him, the wife, and Ms. Hax. This wife will find a hard-working, driven, aggressive, domineering male in her work environment that will sexually please her--and promise babies. This current letter-writing husband will be yesterday's news, despite potentially being a much better mate in today's environment. I suppose that would be difficult advice to write in the newspaper, so I repeat my complete apologies to Ms. Hax. But that's why you have this book—to get the truth you can't get anywhere else. We're not done with Ms. Hax. Just a few short months later; she published the following letter on July 2nd, 2008. Ms. Hax actually does not comment on this letter, which I really regret.

By now, you should already see the outline of the *Peace Machine* process in your daily lives. The woman in the letter below has gone one step beyond that—and cracked the whole thing. As I said in the beginning of this book, these philosophies only seem complex. As the woman in the next letter to Ms. Hax demonstrates, it's actually very simple and very obvious. We all are just in denial about its existence.

> By Carolyn Hax
> Wednesday, July 2, 2008;
>
> While I'm away, readers give the advice:
>
> On marriage and diminished sex drives:
>
> I am a well-off, professional, 38-year-old woman, married 10 years to another professional, with two small children. I love and respect my husband and our life together. I have

little to no sex drive, and at this point, feel like I could happily go the rest of my life without having sex. Is it unreasonable to request that advice columnists, doctors, therapists and husbands acknowledge that decreased libido can be a natural development for many emotionally and physically healthy women of a certain age, and instead of trying to get the wife to get horny, to try to get her to see marital sex as one of the many countless acts she performs daily to ensure a happy marriage and household?

I do not believe my lack of sex drive is something to be "fixed" with counseling, self-help books, drugs or hormones. I am emotionally and physically healthy, have no relationship problems and am not a victim of abuse. While recognizing that these stated issues may be the cause of a diminished sex drive for some women, I propose that the run-of-the-mill decrease in libido that occurs in a married woman in her 30s or 40s with young children is biological, physiological, hormonal and makes sense from an evolutionary standpoint. I thoroughly resent the suggestion that I should endeavor to force myself to desire sex against all of my emotional and physical impulses.

The reality is that it has become politically incorrect to tell a woman it is her "wifely duty" to have sex with her husband. The thought of coercing a woman to engage in sex when she does not want to is unappealing, but that is the issue here. So instead, we avoid the real issue by suggesting methods of resurrecting a woman's sexual desire, which frankly seems ridiculous.

A more thoughtful and helpful method of addressing this important issue is to ask the question, "What do spouses owe each other?"

I don't know if I "owe" my husband regular sex, but I do believe that I owe him certain things, including the ability to compromise, to meet halfway and ensure his happiness to the best of my ability.

My husband owes me a rational, reasonable expectation of sex (let's say, less frequent than every night, more frequent than four times per month). We have regular intercourse, and I'm a good sport about it, because I do believe it is part of my, as anti-feminist as it sounds, "wifely duties." I value the intimacy and fun it brings to our marriage. It is not painful, doesn't take very long and is a small act of giving to further a happy, strong marriage.

I am okay with being told that this is something I should do to maintain a happy marriage. I am not okay with being told that I should see a marriage counselor, buy a book or apply a cream.

Signed; One of Many

The woman who wrote this letter does not even need to bother reading this book. She is in denial in a few little places, but she has the gist of the main ideas nailed dead on. First, she recognizes it's likely that the biological, hormonal, and evolutionary reasons for her decreased libido create a conflict she must solve for the health of the family.

Secondly, she recognizes that the feminist and societal lies, usually told in columns like this, (again, no slight to Ms. Hax intended here, because this is a societal lie) have made it hard for her, or anyone today, to rationally discuss this issue *scientifically*. Our C-brain lies have become societal lies, which we then accept as truths.

This woman owns the secret to success. I mean that because this woman is highly successful and in all likelihood will raise healthy children *and* hold onto her husband. She is not ashamed to admit keeping him is a positive goal--good for her and the children. I'm not saying the reverse should not be just as true and obvious to him. However, for the reasons discussed in this book, it generally won't be as obvious to a man unless the man is motivated by love, something that occurs less frequently in men.

The woman who wrote this letter is motivated by love. She loves her children, loves her husband, and trusts him. She knows he isn't faking his more urgent sex drive, and knows her non-existent one isn't due to illness or a mental problem because she is smart enough to reject those lies and look at her own body for what it is—a miraculous machine that has a reason for everything it does.

Honestly chart the sex drive of most really happy middle-aged women with children at home—get beyond all the lies—and you are going to find this woman. She clearly signaled an understanding of this, in her sign-off. If we motivate our actions by love, as this woman clearly does, we can't be fooled by the lies our mind tells us, or the ones feminists, chauvinists, or society does.

> Don't think right or wrong about men *vs.* women issues. Don't think suppression *vs.* equality either. Think <u>science and survival</u>.

This manner of thinking will cause life to make more sense. If we want to be happy, we need to do what this woman did. Figure out what will make us happy as a family and then construct a life to deliver this goal. If someone tells you to act a certain way—and that action is either unhealthy or not lovingly motivated toward you or your loved ones—don't listen.

Now let's mix things up. What if this terrific woman's husband had an affair despite her obvious reasonableness, intelligence, and loving attitude—and left her? This is possible because of how overly D-brain dominant men are for much of their lives.

While a woman with this attitude is exactly the type that has the best chance of staying happily married, the male's desire to spread seed is motivated at the D-brain level, and it can rear its ugly adulterous head even without intra-couple provocation. If her husband sexually strayed, what would happen? I believe that this woman, with her D-brain love motivation and C-brain sensibilities, would act in the children's behalf (and her own) and resist the temptation to make war.

For females that choose manipulation at the D-brain level, the likely answer would be different. The D-brain would want a divorce, the nastier the better. It would want revenge. It would want him to suffer and **would be willing to use the kids as leverage to punish him**. A destructive overreaction in a C-brain dominated community that can squander precious resources.

Ancient males were programmed to mate with multiple females. If one of his mates burns up all the deer meat in the cave and destroys his relationship with the children because of this extra-relationship mating, she will suffer throughout the rest of her life--and more importantly, so will the children. Quite frankly, he probably will not. She should forgive him in troubled times, and accommodate him in the good ones. And vice versa, of course.

I know that statement will not be popular because as the woman above explains better than I can—it's just not politically correct to discuss "wifely duties," and "what spouses owe each other."
We have a false bias because we see relationships from the C-brain point of view taught to us by society. That view is so filled with the lies we've told each other for generations that it misleads our C-brain into actions that are not in our best interest. If we stand back and look at this scenario from an

evolutionary and objective viewpoint—we get different answers about what our conduct should be.

I'll give you one more example—but this time *not* in the sexual arena. **Still, the brain circuitry/logic is identical.** Remember that point. Whom you mate and whom you hate is more a function of inherited brain programming than we are willing to admit.

At the end of the next chapter I want you to put this book down for a few days, before returning to it. Examine your life and ask yourself honestly if you've thought or even said something racist.

Now, are you a racist?

You probably *can* remember having said or thought something racist but you probably answered no, you're *not* one. Sorry, but you are a racist, whether you are white, black or yellow.

First time readers please go to the next chapter. If this your second read through, please study the following Addendum.

Addendum...Optional Reading

The section below is *optional* reading before the Racism chapter.

It is a glimpse of universe creation and the laws of physics concerning that event which we will explore in logical sequence in later sections. It's presented here in a shortened format because the C & D brain discussion is not finished without including some of this information.

Our brain, and the way the universe was created, though seemingly unrelated, are actually tied together. And since what ties them together is God, or the Singularity, or the black hole that created us, this unification is extensive.

I felt it important that these topics be inserted here to allow the second time reader the opportunity to make the connections between interpersonal relationships, the laws of physics and the creation of our side of the universe.

The simple truth is, all was created by the Singularity, the One (1), God, the black hole, etc. Therefore these principals and laws extend through everything. Do not expect to fully grasp this idea the first time through, hence our advice for first-time readers to skip this section. Studying this from a few different angles, time, perspective, and some repetitive effort will allow you to finally get it.

How is the creation of the universe tied to the decisions I make concerning my marriage?

> It's the motivation of our D-brain--either love or self-interest--that is at the nexus of every decision you make in your life regardless of why *you think* you make them in your conscious mind. This concept is at the very root of how the universe we live in was created, and God's role in it. He is in touch with that decision making process in each and every human. In fact, as we will discover in the Unified Field chapter, He is aware of every atom on our side of the universe.

God is not a third party observer. That idea is mis-leading. *Please* be careful with your thinking here. Adding a judge, our current conception of God, creates a third point of view: the observed, the observer, and the judge. This idea is wrong. Our universe is a two variable universe; we have only two points of observation available. Your brain hurts as you contemplate infinity; this is the barrier you are up against. The Singularity is infinite and is the first variable. The point-of-view-of-God, Love, black holes, the Singularity--this group of things (which are actually one), occupy the slot of the first variable.

> This leaves human consciousness as the second variable.

That's why our universe has the Uncertainty Principle built into it. We represent a second point of view. If a person observes an atom, he changes that atom by virtue of that observation. This is a commonly understood principle of Quantum Mechanics. I believe it is accurate. Therefore, doesn't the atom's presence (God's observation of the person) also change the person?

So, for Descartes and Berkley, let's change the question/challenge. Instead of asking: *"If a tree falls in the forest, and no one hears it fall, did it make a sound/fall/exist?"*

A better question would be: *"What is math?"*

The answer to those two questions is one and the same. Math isn't just an equation that must balance; math is a description of the rules of the universe. Math is a description of a problem and/or question and its solution. This means problems/questions and their solutions exist.

At infinity, no solutions are available *except those that include infinity*. Therefore, the Singularity does not need math. The creation of a second point of view, in addition to the creation of space and time, necessitates the creation of math as a description of what occurs in that universe. Mathematics are instructions for *the second point of view*. No second point of view, no need for math and no free will for mankind. There would be no choices welling up in our D- and C-brains--offering us God's thoughts--the choice motivated by love. Math is the blueprint designed by God's universe (the SU) for the governance of the universe we live in, the Non-Singularity Universe or NSU. In other words, math describes where the C- & D-brain, physics, God, the Uncertainty Principle, Jesus, Buddha, Muhammad and Gandhi, etc., all intersect.

Why is this material at this place in the book? Because our universal decision making process uses the C- and D-brain relationship to offer humans that second variable.

The output of this relationship is the basis for motivation, emotion, and creative thought. When you are motivated to do the loving thing, that urge from your brain is God offering you a preferred choice. He has full knowledge of your circumstance at that instant in time. Your C-brain can then agree, veto, or modify that urged option and create a thought, action, or emotion. Only a Consciousness that knows your motivation, **all** of your circumstances, and has supplied a correct answer that you could use–is able to judge your decision. *No problem can be solved completely without accounting for every variable.*

The C- & D-brains exist inside your brain because of your physical evolution from a creature that chose love as an evolutionary adaptation. None of this contradicts the physics **or** logic of our universe. How does our understanding of the interplay between the C- and D-brain help us understand our relationship with God? This varies from person to person, but for most people it can be as subtle as the difference between the two alternatives that spring to mind when faced with even a simple choice.

If the choice is stark, like hitting the boyfriend or kissing him, both can originate from the D-brain. But, if it's a choice between a playful kiss, or a threatening shove -- a choice between social manipulation and aggression--it is truly the C- and D-brain offering alternatives.

Another view at how this actually plays out in the brain would be described this way. An urge to do something originates in the D-brain. This urge moves up to the C-brain level. The C-brain then has three choices. It can go along with the D-brain; and, these C- & D-brain agreements are perceived as strong will. Alternatively, it can disagree but cooperate, creating a C-brain rationalization. Third, it can attempt to overrule the original D-brain idea--a stressful choice.

The C-brain makes these choices based upon what it has learned from experience as well as the C-brain patterns chosen earlier in life. Additionally, input is provided silently according to the nature of the D-brain beneath. Here's the really exciting part!

Even though the D-brain initiates all the decisions and has the majority of the power, the C-brain can change the D-brain. These are not usually huge changes. They occur little by little through conscious effort. The changes are made physically in the brain as well. This is called Neural Plasticity.

Current scientific thought on the brain is centered on the idea that the brain doesn't change much after maturity, and is just a physical organ. It is believed that personalities come from neuro-chemicals and genes. This is only half true. The interesting part is that what you want the brain to become, it will move toward physically and chemically. Somehow, that effort becomes reflected in your genes passed down the line. I know this flies in the face of modern genetic beliefs, but it is true. In fact, for the last hundred and fifty years a French scientist, Jean-Baptiste Lamarck was ridiculed for asserting this. Ironically, Darwin himself thought Lamarck quite bright.

This huge scientific principle was proved in one simple experiment. If you starve rats, their children will be born smaller. Feed those offspring well and their children will still be smaller than normal. This is not possible unless Lamarck was right and everyone else wrong. According to the Revelation, as described earlier, this makes a gigantic difference in the ways things work and a big difference in evolutionary outcomes.

So, the only way to change the nature of your D-brain is through a conscious effort to change the motivations for your behavior as well as the actual behavior itself. Ultimately, you can only pick from two choices, a love-based choice or any other option. This is the truth of Free Will.

One of those choices is God's choice, communicated to each conscious mind in the universe. God's choice is the one motivated by love. The other choice is motivated by eons of D-Brain evolution or C-brain rules and rationalizations: whichever is dominant at the moment of the decision.

A definition of love would be very useful here; but it's hard to explain.

One parent spanking his child looks exactly like another. Yet one may be motivated by love for the child—the parent is concerned the child may return to play in the street if not punished. The other has the same excuse but was actually motivated by anger over something else. The spanking by the "bad" parent could just as easily be defended, but they aren't the same. Not to God, and not to the child: both know better.

The choice one selects, good or bad, is more likely to be selected again. Behavior is therefore largely patterned, right at the synapse level. BUT, because of this, behavior is not too difficult to change. All it requires is the desire (or will) to change the pattern. This motivation to change must be genuine. The laws of physics demand that brain change come from the creation of a nexus (some people would call this concentration) of the brain's power on a choice. This requires the use of energy, and therefore change cannot be initiated any other way. This is the power behind Cognitive Behavioral Therapy.

Brains don't change without effort, concentration, will, desire, and choice. These concepts, taken together, really just represent extra effort. The current thoughts and behaviors that occur in the normal course of events constitute a type of pattern. Patterns are what brains are great at executing without a lot of energy. This lowers the cost of the operation from an energy or calorie perspective -- an efficiency move. Your brain executes patterns without much effort, but can physically change those patterns through repetition and concentration/extra effort.

C-brains literally wire themselves from experience, strengthening those connections when there is repetition. They re-wire themselves when necessary. The accumulation of these choices contributes to the decision making process for the next D-brain urge racing up out of the lower brain to be challenged (or not) by the cortical brain, the C-brain.

The ramifications of this setup taken together with the Unified Field Theory, presented later, are staggering! Each human can re-wire his or her brain (gradually) to make it into whatever you want to be. God is leaving us instructions every single day, concerning every single thought, about which path to take. These "instructions" are in the love-motivated option that seems to come out of nowhere in our thinking/brain. Choose it consistently and your life changes.

The brain's behavior in this way is not the result of a miracle. It is why the brain had success as an evolutionary agent for millions of years. We are the current end product of that process on the planet Earth. Our brains are the cumulative result of a huge evolutionary laboratory.

Past decisions become pivotal because they affect which brain side is in control during the execution of future choices. Repeated choices become patterns. This is basic to our evolutionary blueprint: to get guidance to enhance our survival, and also to **impart feedback data about choices made (and not made) into our genetic record.**

The structure of the universe, the structure of evolution, the nature of free will, the presence of God in our universe, are all inter-related and echoed at the physical level in the human brain. The brain's most fundamental operation is the day-in and day-out execution of the decision-making process involving competing motivations.

Every choice. Every day.

Book I, Chapter 5

Racism I

We are all racists in the D-brain. Some of us are racist in the C-brain as well. Most liberals believe everybody is born without prejudice, that racism is learned. The reverse is true. The D-brain is naturally programmed to fear facial color and even some facial features different from its own. Because of its central importance in the Revelation, we will return to this topic at the conclusion of Book II in a second racism chapter. This chapter lays the groundwork.

Blacks warred for generations with whites and browns and yellows, and vice versa, tribe-by-tribe, island-by-island, people-by-people. Most races/tribes were constantly at war with other humans. Most used facial (skin *color* tops this list) and other visual recognition clues to determine friend from foe for extended generations. This mechanism is deeply ingrained in our D-brain. The article below is just one example, from among hundreds available, on the power of this process.

> **Aggression written in the shape of a man's face**
> NewScientist.com news service
> Gursharan Randhawa
>
> No matter how hard men try, they may not be able to hide their aggression. A study in male ice-hockey players suggests that to gauge a man's aggression levels, you just have to look at the proportions of his face.
>
> Cheryl McCormick and Justin Carre from Brock University in Ontario, Canada, found that the larger the width-to-height ratio of a player's face, the more aggressive they were. They measured aggression by the number of penalty points each player accrued for potentially harmful behaviour, such as elbowing and fighting.
>
> In general, men's faces tend to have a larger width-to-height ratio than women's. This physical characteristic has been linked to higher levels of testosterone, which in turn is linked to aggressive behaviour.
>
> Most people would not want to pick a fight with a big, brawny man, but because facial ratio is not linked to body size, it may have been favoured by evolution to warn others of an aggressive personality they would not want to tangle with.

Real world results

Although the team first found the result in a study of students playing computer games, McCormick says they were "astounded to see that the measure could predict aggressive behaviour in a 'real world' setting".

Previous studies on facial metrics have suggested that women can tell whether a man wants children just by looking at their photograph. Now McCormick's study raises the question of whether people can spot these subtle facial differences and use it to guide everyday behaviour.

"If someone was given the choice of one of two opponents to compete against who differed on the basis of the facial metric, would the facial metric predict the less aggressive opponent?" asks McCormick.

She believes that people's faces may be influencing who we chose to socialise with on a daily basis.

Reference: *Proceedings of the Royal Society B*

The D-brain uses skin color and facial shape because the D-brain gets direct visual input and has million of years of genetically stored memory to compare those images against. The D-brain is visually influenced to a great degree. Remember the brain's massive connections with the optic nerve complex. It prioritizes this one sense over all others. Humans sacrificed the development of the other senses evolutionarily for the prioritization of sight. But, the sheer volume of data available visually makes it imperative to "encrypt" the data stream traveling through the optic nerve. This is the beginning of conscious thought—a concept I will return to later.

The fear a white man feels when confronting a black man on a poorly-lit street is only matched by the fear the black man feels when he has to confront the white man alone.

Neither race is willing to admit this basic fear exists because of denial and one of its manifestations, pride, even though it is the first reaction people have under most environmental conditions. This reaction occurs even when they have been conditioned *against* being racist by society for years. In fact, as the brain deteriorates over time as in older people, you can chart a pronounced increase in racially-motivated statements in a person's language and behavior.

This is no coincidence. The C-brain ages and weakens first, allowing the primitive D-brain to act with impunity from the C-brain influence. These people become meaner, more aggressive, more racially biased, more stubborn, less controlled sexually, and paranoid. Nurses who have worked in a nursing home will confirm the statements in this paragraph. Often, nurses mistakenly attribute these racist statements to the fact that old people come from a different generation. Actually, this represents a *change* in behavior for most of them. Many non-racist (C-brain) people become racist as they age. They then join the other old people who were racist all along due to upbringing.

Drugs, alcohol, brain trauma, disease, and any other type of lost C-brain control can erode people's ability to control outward manifestations of their D-brain. This can occur even in people that had extensive boundaries, developed over time. Imagine comedian Michael Richards' surprise when they showed the Seinfeld character the video clip of himself repeatedly calling a black man in the nightclub he was performing at a "nigger." I have no doubt that his original statement "I didn't say that" was truthful from his C-brain, *Peace Machine*, memory point of view.

Upon seeing the tape, he must have been curious about where those statements originated. He had no C-brain experiences, feelings, or other racist thoughts that he could summon as a cause for what happened that night. Many of his black friends confirmed that they were as mystified by his behavior as he was. They stated that it contradicted the Michael Richards they knew and liked. Of course, they only knew his C-brain, his social tool, his denial tool, his *Peace Machine*.

The "nigger" stuff came from somewhere else. Richards' D-brain genetic patterning "remembers," black skin equals confrontation and possible death. The D-brain got to "talk" that night because he was either intoxicated or high. Intoxicating substances lower the C-brain's respective power and control over D-brain urges. I could cite a few *thousand* other examples that fit this pattern but I won't – you know too many already.

Now I challenge you to look at race in a new way. You are a D-brain racist whether you are black or white, yellow or tan. You should strive **not to be** a C-brain racist of course. I would add the D-brain to that list but, as you know, I don't believe you have the power to do that alone. Eventually, racism programming will work itself out, because racist behavior is no longer good for the species. The D-brain adapts--slowly. Since your D-brain will not change today or tomorrow, you'll live with these negative instincts better during your lifetime if you can recognize their roots, and manage them.

Most white people in the U.S. today think it's time to move on from the race issue. For them, the race issue ended with the Civil War and the Civil Rights movement in the 1960's. Let's examine this from the D-brain point of view. Just a few generations ago, white people sailed boats to Africa and forcibly loaded six million black people onto those boats. They then tortured them, to make them easier to handle, as they transported them to a foreign land filled with more white torturers. This history is not being recounted here to generate sympathy. Two million (one third) of those original six million blacks did not survive the capture and transportation process. Humans are tough--you have to horribly mistreat six million people for two million of them to die that quickly. (Look the numbers up yourself. As always, I urge you to check everything in this book.)

The remaining four million black humans arrived in the U.S., without rights, property, education about how to interact in the new society, or any inherited information about how to survive in this strange land. These types of primitive "subjugating" events have repeatedly occurred over tens of thousands of years, and explain why black people fear white people. It is also why white people fear black people. And why all people fear others who are different from them. Things like this have been happening between different-looking humans for hundreds of thousands of years. Since both black- and white-skin colored people survived, it's my guess that only the recent battles were as lopsided in the white-skinned man's favor as the slave trade turned out to be.

This racial fear cannot be openly expressed as fear; denial and pride would never allow this. The D-brain knows showing fear until the moment it decides to flee is often deadly to the organism. This explains why modern whites, despite their murderous behavior in this particular slaughter, feel the same fear toward blacks.

Our D-brains are not concerned with being "fair"; they just care about survival. If you scare white-skinned humans badly enough, or give them enough economic incentive, they'd kill or enslave again. However, *same* race on *same* race genocide kills *100 times* the number of people today than does different race on different race killing.

The C-brains of a white man and a black man, serving in a foxhole together in a foreign land, can learn to love and trust each other like brothers. It is a bonding that serving in life-or-death situations together can create. But, they still will feel an innate fear toward *other* differently-colored men upon returning home. Consider one last thing: when Hitler had a choice at the end of the war to use the trains to transport supplies to the remaining soldiers defending Germany or transport Jews to the ovens, he choose to transport the Jews.

All the literature we have from this period details the great lengths Hitler and the other Nazi leaders took to categorize the differences in facial features between German Jews and other Germans. They had drawings, charts, and pictures of hooked noses. They created an entire visual catalog of whom to kill. It was an obsessive, scary look at the completely exposed D-brain. Just like the unexposed one you have.

The D-brain's ancient programming is hard to beat because it trumps the C-brain and genetically "remembers" all those millions of years that it used skin color (and facial features) to distinguish who to fear and who not to fear. The evolved brain doesn't lightly discard a targeting-and-recognition system for hatred and violence that has enhanced survival for millions of years.

Now put this book down for a few days at least. **Don't cheat. Stop reading**, I mean it, just for a few days.

Book I, Chapter 6

Denial, Lying, and Guilt

If it hasn't been a few days, put this book back down.

People are good at lying and denial. They do both a lot.

Most people immediately take offense here, so let me clarify. I don't mean lying, like the whopper your Mom spanked you for. It's more like a lie you don't know is a lie. It is the lie you tell:

...Because you were told it as you were growing up.

...Because it was "commonly known."

...Because you want to believe it since it serves your interest to do so.

...Because it is an issue for which your brain is programmed to deny the truth.

The average human brain puts up stiff resistance to most new, different, and, especially, *uncomfortable* ideas that change the "reality" already fabricated in our brains. If this makes any sense at all, you are beginning to understand the world more clearly. Still, it is probably uncomfortable to jettison all the illogical assumptions you previously took for granted.

This situation, coupled with the gross ignorance of our species as we emerged from the jungle, makes our ideas and concepts about human behavior dramatically out-of-sync with our actual experience. This is true physically and metaphysically. This has made it hard for the Truth for a very long time.

The enormous scale of our denial mechanisms, and their long-term effects, are very bad for us individually and as a species. You need to know why, how, and the subjects about which people lie, **if** you want to make choices based on an understanding of the truth.

Your individual life and the world stage around you--institutions, cultural norms, religions, customs, and the laws we live by--reflect our species' unusual brain structure. **Therefore, the principles in this book are just as true when applied to the "world stage" as they are in your individual life.**

The reason I flip back and forth between the big picture and the individual view is that I want the reader to be able to make important connections to his life and the world around him simultaneously. This book is not a scientific text. Instead, it's designed so that you, the reader, can discern these patterns, these underlying truths for yourself.

You may not know exactly how to build an atom bomb, but the knowledge that it is possible is valuable to you as a citizen of a country that possesses such a weapon. Even if you never handle such a weapon personally, understanding societal lies and how they affect our culture, our daily life, our laws, and even whom we fight wars against, can save our individual lives.

Think of a contest where any two competitors--government vs. government, warrior-to-warrior, religion vs. religion, general vs. general--must develop *a strategy for winning* using mostly incorrect information. The poor quality of this information is the result of combining unknown data and misinformation told to the decision makers by humans loyal to the decision maker, who themselves did not know the poor quality of their own data.

Add to this unreliable data from the outright lying done by humans *not loyal* to the decision maker, and you can see how poor the information pool usually is.

The above scenario describes the usual situation in the world. Neither competitor's strategy is based upon sufficient real data. *Therefore, the plans executed by these competitors wouldn't necessarily be based on any underlying truth or facts, and the ultimate reasons for victory or defeat have little to do with information or strategy.*

Consequently, if viewed statistically, the long-term results from these contests/conflicts should be a series of tossups with no clear pattern or winner. This is why empires fall. And indeed, our recorded history *has been* a series of arbitrary, poorly-recorded, bloody, pointless and destructive events – more often decided by weather or chance than by any other factor.

These conflicts usually take the form of war against a foreign people, but include civil war, genocide and riot, by which human world history and

God is love

geography has been decided and power apportioned.

> The stated goals and actions of the human leadership associated with most of these traumatic societal events and a truthful understanding of the causes creating these events lacks any real linkage.

The individual's, and therefore society's, denial mechanisms concerning this fact are the result of a struggle between many different groups:

A. Family
B. Personal friends (or a clan)
C. Political
D. Religious
E. Nationalistic
F. Economic
G. Governmental factions

They are all warring to obtain the authority and/or define what you are told to do. They all want to convince you that what you are *told* by them *is true*. They will do this through whatever information distribution channels are in their control. This isn't a conspiracy *per se*, but all these efforts have a common purpose:

> To aid those already in a superior power arrangement relative to you in maintaining that control longer. Or: to aid those not in a superior power arrangement relative to you, in gaining one.

People in power and the institutions controlled by them are almost universally obsessed with the use, trappings, and display of power. They don't have time to search for God, or a cure for man's miseries, or discover solutions to life's mysteries.

The vast majority of people in power *consider* the common man their personal misery and ailment. We should not therefore be surprised that the efforts of our governmental, scientific, and spiritual "leaders" have not revealed much truth about mankind, the nature of the universe, or spiritual truth.
All people, and I mean all people…including:

The Revelation

1. Family (including parents)
2. Friends
3. Lovers (most miss this at first)
4. Teachers
5. Government
6. Yourself
7. Scientists and doctors
8. Lawyers and judges
9. Politicians

…have been telling you, without consciously lying, things that they didn't *know* to be true. And indeed, most was not true. All of these people had been told unfounded "facts" by people they trusted during the time they were forming their:
1. Minds
2. Emotional responses
3. Family structures
4. Moral and other guidelines.

And their religious beliefs guided by:
1. Family
2. Mentors
3. Priests
4. Mullahs
5. Lovers
6. Children

This isn't a judgment. The important thing is that these people believed that a great deal of the misinformation they passed on to you was true.

It is crucially important to open your mind to the possibility that you, without consciously knowing it, lie to yourself everyday. You deny certain information from having access to your conscious brain without realizing it. "I am not an alcoholic" is just one easily recognizable and frequently repeated example.

Lying and denial are much more important to humans than we currently realize.

Lying and denial work together in the brain in a specific way that has many *important* evolutionary purposes, including the promotion of human socialization. Lying and denial are necessary tools of the *Peace Machine*.

This isn't a defense of lying, nor an attack on it. It's important to remove the moral judgments, which are by definition very unscientific. Remember to keep moral judgment aside as you read on. Our reluctance to face the truths about ourselves has hampered us as a species, and continues to do so.

Communism doesn't work as an economic system because Karl Marx got a basic and critical point wrong. Primitive man **did not work** according to ability and distribute according to need. The underlying premise all his theories were based upon was dead wrong.

Primitive humans used murder, manipulation, lying and brute force to distribute goods and hunting territory -- and we still behave this way.

We succeeded through the use of our evolutionarily honed highly competitive behavior. Hence, when you base an economic system involving humans upon any principle other than competition, it will *always* fail to produce the level of goods and services that a competitive system will. Later in the book, I will explain the critical need for strong anti-trust laws and Market-Fences, because they *protect* competition. We were molded by fierce competition, and therefore still depend upon it for survival.

This is why strong anti-trust legislation, in a Market-Fence model, is the only long-term protection for the continued supply of necessary materials you and your family need: food, energy, and medicine.

Any system that lacks real competition has no checks on the destructive cumulative effects of the individual and societal lying and denial. Anti-competitive, monopolistic, forces are so powerful and prevalent that they completely corrupt productivity through hundreds of "holes' in the economic "gas cloud" (to be explained in the next section), thus eroding cumulative wealth.

Efforts to use laws and policies to control the "bad" human behavior blamed for economic collapses have failed countless times. In fact, they make the situation worse. Humans can lie to justify themselves around laws all day. The Sarbanes-Oxley accounting law was meant to provide additional transparency in corporate America, but fixed nothing. It did, however, further erode the competitive environment in the U.S. by driving and scattering new business competition into different markets. The inevitable result was the damaging of U.S. productivity and the lowering of U.S. standards of living.

Humans are so thoroughly programmed to lie, cheat, steal, and overpower each other that, if not counter-balanced by competition, they can starve themselves to death through lack of productivity/effort. This occurs when one population group attempts to kill another in order to gain control, and then unintentionally destroys the productive mechanisms necessary to support all humans in the area under control. This also occurs in non-life and death situations in the monetized portion of the economy, as will be seen in the next section of this book.

Both sexes practice *actual* lying almost as soon as they can speak. Children do this initially as toddlers, without purpose or reward, *as if it were programmed* into the brain's function. They are just testing/learning their deception abilities. The parts of the human brain responsible for denial and lying have sectional layers where the truth is allowed to penetrate. But certain behaviors, thoughts, actions, and words are out of reach of the conscious mind reading this sentence and will remain so, absent a conscious effort to learn or re-program.

Mass denial, mass hysteria, and irrational hatred of other humans because of color, race, religion, or <u>even</u> socio-economic class, all co-exist within humans who consciously loathe that kind of behavior in others. That is the true norm, not the exception, even though virtually no one will admit it. This is why humans *needed* the *Peace Machine* in order to successfully live together in groups.

This psychological dichotomy is a successful adaptation based upon our physiological needs as viewed from an evolutionary basis. This book will show you how to turn that from a disadvantage into an advantage. With information about our brain and thought structures, you can make your brain work much better for you than it currently does.

People lie to themselves so often, that the respected scientist, Michael Crichton, feels almost any conclusion found by **any** group of investigators or study authors is bound to have the conclusion affected by the inherent bias— the original belief of the investigators. Mr. Crichton did specifically exempt double blind studies that use placebos from this conclusion.

A placebo is a "fake" medicine (or sugar pill) that you think can help you— and then it does somehow. The more interesting question then becomes: Why is there a "placebo effect" at all? It's proof that people can change reality in their body with a false belief. The placebo effect points to the power of belief, even false belief, in humans. **This helps to explain why the power of belief itself, is such a successful evolutionary adaptation.** Lying, denial, and guilt run deep in the brain.

A Word or Two About Guilt and Self-Awareness

Classic guilt is a C-brain emotional constriction of the blood supply throughout the body, a physical reaction designed to warn the D-brain that it is getting the C-brain upset. Though the C-brain believes that it is in complete control of the body, it really isn't. How can this have happened?

While the C-brain controls much routine human behavior in a reading and writing world, developing habits and behaviors, it rarely realizes that the D-brain is capable of adultery and violent behaviors including murder.

There are many D-brain behaviors that the C-brain consciously knows are wrong, illegal, immoral, unethical, prohibited, and/or forbidden in society. The C-brain is generally unaware of its D-brain willingness to do these things, until the D-brain does them.

The guilt-racked C-brain is literally nauseous over the initial realization that it doesn't really know the whole person it represents. "I don't approve of this behavior" the Conscious mind screams—"What part of my brain possessed me to do/say/act this way?"

Every C-brain carries an image of itself: *"a perceptual picture of me."* But, the C-brain was created as a tool by the D-brain to enable it to lie to other humans about its ultimate intentions while attempting to simultaneously exploit them for sex, protection, labor, deer meat, or hunting help. This is the true nature of our "socialization" tool.

Unfortunately, the brain image we rely upon for our internal self-portrait is, *at best*, only half the truth! The other half of the truth comes from your D-brain, which is *comfortable* with a spouse, three kids, a full-time paramour, random sexual relationships, and taking your neighbor's wallet/deer meat and wife.

A person's ultimate behavior is a collage of C- and D-brain decisions that choose things like sexual partners, where a heavy percentage of D-brain impulses get obeyed, or performing math and geometry to design a machine or building, where a heavy percentage of C-brain impulses get obeyed. The two sides may, in certain instances, even share their opinions of each other.

The often-opposing forces for thoughts and

motivations, while providing us with the opportunity of choice, also provide fertile ground for internal conflict that can have destructive effects on our emotional and physical health. So the question becomes: how is this an evolutionary advantage given the risks involved?

The C-brain has a probability cloud of behaviors that it considers bad for the health of the organism, or bad for ethical or moral reasons it learned. This C-brain cloud is denser in some areas than in others as regards the probability of a particular behavior or action resulting in death or harm.

Cumulatively, the risk of death is in direct proportion to the number of times the brain selects the more risky behaviors. So, something risky can be done successfully, but usually only a small number of times or by a small number of organisms. Conversely, even the safe decisions *will fail* a very small number of times.

There is no guarantee of complete safety even by choosing to live exclusively in the safest part of the cloud. The large number of choices makes sure there are a wide variety of behaviors exhibited by the organism's *population* as a whole. This ensures that even the weirdest behavioral strategy has a chance, no matter how minute, to contribute to the survival of the species.

Humans had an explosion of success through use of a very "weird" or uncommon adaptation.

The dinosaurs were the final word on the big, strong, ferocious strategy. But that strategy had a flaw: it can't ward off asteroids or ice ages.

Dinosaurs, the furthest evolutionary extension of the "Big" Fundamental, lost out completely. A lot more sea life survived using that same strategy (Big) because they live in the ocean, where animals bigger then a vole *did* survive the asteroid and ice age threats.

The vast majority of organisms will execute the safer probability paths, and have the highest chance of passing on their genes.

Guilt ultimately works out to be a survival-enhancing mechanism by steering us to the safer areas of the probability cloud. Guilt is also the remorse our brain feels, mostly by the C-brain, when it ignores the love-motivated option.

Empathy

However, we can feel guilt for circumstances we didn't create or couldn't avoid. This guilt is empathetic in nature, love-motivated. Empathy is healthy for the wronged and the empathetic. Our D-brain remains highly skeptical of empathic feelings. It leaves this emotional calculation to its creation, the C-brain, the layer of the brain that allows us to socialize, mate, cooperate, and collaborate in the first place.

Guilt, then, serves multiple important evolutionary purposes. It allows a punishment system to work between the two layers of the brain while maintaining the ability of denial for the purpose of unified command, which, as already discussed, is critical for survival.

As negative as all this sounds, don't get discouraged. There is a safeguard in the system, which has gotten us this far as a species, serving as a counter-weight to all this negativity. In fact, objective learning about how the *Peace Machine* works makes it possible for us to leverage this great counterweight —advanced-nurturing and love-motivated-decisions--to benefit ourselves and mankind.

The Revelation

Book I, Chapter 7

Abortion, Privacy and The Constitution

Abortion is a classic example of how the two parts of the brain work together *and* work against each other. It is also a classic example of how the laws and rules in society conform to the C-brain, but only after it gains dominance in the majority of the population. This chapter is neither a tirade against abortion nor an argument for it.

The D-brain of a woman:
Loves babies.
Likes sex.
Likes to be dominated.
Is very, very, practical.
Suffers little to no damage from an aborted pregnancy of which the D-brain disapproves. However, can suffer serious damage to the C-brain, done by the D-brain, if the abortion involves the loss of a fetus the D-brain was content bearing.

Please re-read that last point above. Most people have such a large emotional response to this issue they cannot read it objectively. Most automatically form objections to varying opinions about the abortion topic so quickly they cannot evaluate or absorb new data. The brain system described in this book does not mean *every* abortion hurts every woman's brain. Women who have had the procedure, but later on raise children—are also largely protected from this brain-reaction.

Alternatively, the D-brain may know something's wrong with the fetus, or estimates that there are not enough resources available to keep itself and the child alive, thus accepting abortion with minimal trauma. Women have these types of protective safety valves for sound evolutionary reasons.

Why does this brain safety valve exist? Because having babies, until the recent invention of forceps and safe C-sections, was difficult and dangerous. If the loss of a stillborn or miscarried child incapacitated a woman too much emotionally, she was less valuable from an evolutionary standpoint; she could not care for her other children or attempt to bear again and provide support for her tribe/family.

Therefore, there is a mental safety valve to prevent damage to the brain over an "evolutionarily understandable" loss. **But** it only protects women who lose or purposely terminate a baby under certain circumstances.

> The primary purpose of the female D-brain is to bear and raise children, whether or not the female C-brain in question recognizes this. Where the C-brain doesn't recognize this basic truth, expect a high wall of denial.

So, if the D-brain is counting on a baby (it knows it's there) and finds out the C-brain had it sucked out—the D-brain often takes retribution against the C-brain, resulting in mental illness such as mood disorders, depression, suicide, and drug use. Since some of these psychological responses are delayed, it can be hard for people to pinpoint the cause of the mental illness. Middle-aged women in the U.S., from 26 to 49 years of age, have experienced a doubling of the suicide rate in just the past ten years--and this is part of the cause. The other contributing factor is the aforementioned conflict between modern females C-brain goals and D-brain goals.

Society would talk about these women and these problems more except for the C-brain denial issue. The media has been very C-brain aloof on this subject, and has done virtually no investigation in to what damage abortions do to women that have not had a child previously or subsequently. Before we narrow our focus too sharply, let me interject that abortion is only the most obvious triggering device here, not the sole cause of emotional trauma and malady for the modern female.

> The difficult struggle inside the modern female mind between what her D-brain expects of her and her mate, and what her C-brain thinks she and her mate should do, occurs in *almost all modern industrialized societies except those that make great efforts to accommodate, protect, encourage, and support the modern woman in her dual roles*--mostly socialist countries like Denmark. Consequently, the modern C/D-brain cross-purposes conflict is damaging many women who have never even had an abortion.

An abundance of psychotherapy terms exist for the multitude of disorders that afflict the mental state of women in the U.S. But all told, the reality is this: a huge number of American women are suffering from diagnosed and undiagnosed mental disorders related to D- vs. C-brain conflicts they suffered over the last three decades. The percentage of mental disorders in otherwise healthy, intelligent, educated, economically successful women in metropolitan areas in this and other highly developed countries is staggering. I don't say this because I am against abortion or "have it in" for feminist women. On the contrary, I have argued for abortion rights **publicly**, a number of times. I even editorialized in favor of it, prior to the Revelation. I would not have done this if I had not been convinced that abortion was not as dangerous as I now know it to be. This problem is readily apparent based upon the structure of the *Peace Machine* in the Revelation. It is confirmed upon observation. It is a truth, whether we like it or not. I am not judging behavior or setting morality.

The C-brain of every woman, just like every man, thinks it's in charge. Here is where all the information a woman has learned throughout her life is stored, every crazy conflicting idea, from using certain chemicals on the hair to change its color, wave, body, bounce, length, and sparkle--to ideas about what causes cancer and who should have won on "American Idol."

The conscious mind of even the most distinguished female thinks about her hair more than men do about sex. This is **not** a put down. Women concentrate on their looks because their D-brains, from generations of evolution, know men use their eyes to choose a mate. This confers upon the most attractive female elevated social status, as they are most likely to be chosen as a mate by a leader/hunter, or to *become* an actual leader/hunter themselves in the modern environment. Human C-brains go along with this, though they do not know why.

Caring about their looks is matter of survival for female D-brains, and female brains know it. Why else would the fashion/grooming business be such a big money industry for females and not nearly so for heterosexual males? It's a D-brain imperative for women. Attractive women understand this situation. They feel stronger, more powerful, and dominant when they think they look good, whether or not they are willing to admit it. In the same way that hunting/working provides men with a slight emotional buzz, having their hair worked on, their nails done, etc., provides the healthy female mind with the same buzz.

Generally, the female C-brain is also religious. Seventy-five to ninety percent of American women consistently profess a belief in God. This should be the end of the abortion question. It is not, because of mass denial. If women actually voted religiously at the ballot box, abortion wouldn't be legal. You need the power of mass denial to make this statistical paradox understandable. Seventy-five percent of the female populace should have enough political power to outlaw something in power-sharing societies. However, religious belief is selective in the human C-brain.

The *Peace Machine* part of the female brain believes in the parts of the Bible that make sense to it, like forgiveness -- but not the parts about multiple wives, stoning people, the wrath of God, and male primacy. This is expected because religious beliefs can only be selectively believed in a society that has denial built into its logic system.

The female D-brain believes in the Golden Rule: "Thou shall do unto others, as you would have done unto you." This is a D-brain inspired belief because it is motivated by fear, the strongest D-brain force. The Golden Rule to the female D-brain really reads like this: "Thou shall not hurt others too badly because it will make them angry, and they may come to burn your fields, slaughter your husband, sell your children into slavery, and then rape and kill you."

Since many D-brain-heavy men *enjoy* these activities, males are less concerned with the Golden Rule. Men are the killers the female D-brain fears. Men's D-brains believe "He who owns the gold, or is the strongest, makes the rules." The female C-brain believes what it is told by doctors, lawyers, and other professionals...but not as much as it once did. It believes what Mom told it, and still believes her more than anyone else. Family relationships are D-brain controlled, especially child to mother.

Many females continue strong relationships with their mother -- decades after these relationships have become highly damaging and destructive. The D-brain rules here as well. Just because your C-brain, your psychologist, and your husband all agree that a dysfunctional relationship with your mom is unhealthy, many women cannot change the influence their mothers' opinion has on their mental health, their decision making, and their relationship(s) with others.

Remember the power of the D-brain; it trumps all—without recognition or understanding. The female C-brain walks a fine line in modern power-sharing societies. It makes every decision based upon trying to balance what it believes is right, what will protect/benefit her or her family, and what the official/societal rules say should be done. When really confused, it may ask its mate (or more likely, mom) what to do.

Asking another person what to do is a female C-brain favorite alternative that is not generally employed by the male C- or D-brain. Men make a lot of stupid mistakes for this reason. Male D-brains don't like to ask advice; it represents hesitation and weakness to the hunter/killer, and is therefore judged too risky by male D-brains. Most women should recognize their husbands here.

In reality, neither the male nor the female C-brain is really making most decisions. The C-brain is just supplying the rationalization needed by the D-brain so it can do what it wants. This brings us back to the emotionally-charged topic of abortion: originally illegal, then permitted state by state until finally legalized everywhere by Roe v. Wade, a 1972 Supreme Court decision holding that the Constitutional protection of privacy should be extended to include a women's right to terminate a pregnancy. This is the worst Supreme Court ruling in U.S. history with the exception of the so-called "Dred Scott" ruling.

Neither existing Federal law, nor the wording *or* drafting of the Constitution actually supports extending the Constitution's non-existent privacy protection into this other (abortion) non-existent area. The Supreme Court is supposed to rely on Congress to make the laws and then decide adherence to the Constitution on the basis of pre-existing Federal and Constitutional law. And yet, in this particular situation, the Justices did not. Oddly, the ruling itself doesn't address any of these core issues. More curiously, it fails to address any other critically connected abortion issues such as age, parental notification, etc. Most amazingly, the Roe v. Wade ruling literally prohibits *Congress from enacting any future laws, which prohibit abortion or legislate the issue or any of its peripheral points.* I wonder how many Americans actually understand that last part?

If any other new medical procedure appears in America tomorrow, such as safe abortion procedures did, would the Supreme Court insist that some vague area of the existing Constitution applies to define this right? Or, would they simply ask our legislators to make a new law or even offer a new constitutional amendment—if they decided it was appropriate or needed? Even if they made a wrong choice and decided to make up law out of thin air, how can it be explained that they then blocked *all future modification of this law* by the Congress, the President, and therefore the public?

Clearly, the answer hinges on this new "right" being protected by the D-brain of the male and the female. This situation also explains why this legal embarrassment has stood all these years. The closest thing to the faulty logic the Supreme Court used in Roe *v.* Wade was the argument employed in both Dred Scott and Plessey *v.* Ferguson, two other rulings based upon emotion and not constitutional law or logic.

I don't point out all these problems with Roe *v.* Wade because I oppose it, but because the problems themselves point to the fact that its creation was clearly motivated by forces stronger than the system itself. This ruling is the story of the C-brain and D-brain in America during this time period.

Laws reflect the personality of the national consciousness, which in turn reflect the C- and D-brain conflicts in all of us -- collectively. Men poll 50/50 on the abortion issue because they aren't staked through the brain by it like women. An abortion alters a woman's brain chemistry, not the males. As a result, women vote overwhelmingly for abortion in a private ballot *but not* in open elections or polling. This indicates the female brain wants the right to get out of a bad relationship instead of being trapped by childbirth, even when a women's conscious mind, her religious mind, supports adoption instead, for example.

If men actually did vote 50/50 this would reduce their voice to nothing, and as a result women would completely decide this issue. But men also lie about this issue, and privately vote for pro-choice candidates in bigger numbers than they openly poll for it. They do this for the same reason women do, out of a fear of being trapped with a woman they otherwise might reject but have made pregnant. The entire Supreme Court writing Roe v. Wade was male.

Of course, Liberals (high C-brain dominance) are for abortion rights. But they can't carry the day without women who otherwise would be expected to vote "no" on this issue - the religious women who generally vote conservatively on other issues.

The public wanted a Supreme Court ruling. It protected their desire to have this option without forcing them to vote for it openly. This is not hypocrisy, but rather C-brain denial. The belief in the Supreme Court opinion occurred simultaneously in *millions* of women, or the current situation would not be possible mathematically.

You cannot have majorities of people voting for two sides of an issue. Which proves the existence of mass denial on the abortion issue. How would you attempt to convince somebody to vote against something they consciously agree with you is bad—but privately protect in the ballot box—an issue they protect even from self-admission?

A million babies a year are aborted in the U.S.-- approximately 250 to 300 for every 1000 live births, or more than one out every four babies. Half of all pregnancies are reported to be, unwanted, accidental, or unplanned. Think about that number and how these women feel. You can practically **feel their panic or joy** at the moment of discovery--depending upon their domestic situation.

That's why feminists support abortion rights so staunchly. They believe women must have an option to prevent them from being forced into a corner that would potentially affect them the rest of their lives. What they don't know medically, spiritually, and brain structure-wise is that the exercise of the abortion option, in certain circumstances, can have disastrous life-long effects on the mental health of the woman.

American women only plan or want 50% of the pregnancies they experience, so they want the right to have an abortion. They vote for pro-choice candidates, support presidential candidates that will nominate the "right" people to the Supreme Court to protect it, and prove their allegiance to this procedure by using that right one million times a year.

No one ever openly *supports* an abortion; they support the choice to have one. But it's like slavery: you can walk among it, live among it, have wives among it, even write great historical documents to abolish it--while still practicing it. How? If it is something that your C-brain rationalizes it has the "right" to do, and your D-brain fears it can be trapped without that *option*, the brain will go ahead and do it, despite obvious contradictions.

While this may sound harsh, it is true. If something sounds harsh and is true-- we normally try not to think about it. However, it's much more important to understand something we don't want to talk about than something we all agree makes sense. Otherwise, serious C- vs. D-brain conflicts arise that eat away at the roots of individual sanity, harming society.

For this reason, society needs to address the side of the issue represented by the women who had abortions and are now suffering the emotional consequences--serious mental difficulties ranging from borderline personality disorder to a range of "mood disorders" including drug and alcohol addiction, depression, and anxiety--while in many cases *consciously* being satisfied with their abortion decision.

Even in the woman who chooses to abort, the D-brain, the stronger side of the human brain, wants a baby to love and nurture. The D-brain knows which brain is responsible for the problem when that doesn't happen. This is bad news for these women because D-brains *are* ruthless. The female D-brain will kill in self-defense. Unfortunately for these women, it often interprets abortion from that point of view. And, their D-brains get revenge.

A C-brain killer profile paired with D-brain killer programming happens much more frequently in the male brain, where such purity of purpose is rewarded in the hunter/predator role.

This combination of C & D killers is not generally coupled in the female brain because it creates too many conflicts of interest in the nesting and baby producing females—their *primary purpose* by evolutionary rule.

This chapter is not presented to dehumanize this struggle or its effects—that is not my intent. On the contrary, it's easier to understand the significance of something you deny with the C-brain side, if you can also see it from the more powerful D-brain's point of view.

An abortion will have an awful impact on the brain of any female from *any animal species* for whom producing and nurturing offspring is paramount. You can't have it both ways. We can't love, sacrifice for, and receive happiness from children at the core level of our brains, while also believing that it *means nothing to have them destroyed*. We must learn as a species that our C- and D-brain structure makes this conflict and others possible today. The effects are almost never reported because of individual and societal denial. Alternatively, we can continue in denial and have no hope of overcoming the damage we do to ourselves through abortion, racism, nationalism, and religious and ideological fanaticism. All of these ideas are "cover material" in the C-brain fabricated to protect the institutionalized programmed killing the D-brain wants to do, or, in most cases of abortion, *doesn't want to do.*

Any alcoholic will tell you, the first step to sobriety is admitting you have a problem. If we are going to allow abortion to continue as a choice in our society, we are obligated to provide the medical, emotional, societal, and psychological support necessary to prevent the enormous brain damage it does.

Book I, Chapter 8

Personal Conduct

1. Always act/negotiate/demonstrate in good faith.

Since virtually no one else does, this may seem silly. But remember: the vast majority of negotiations and actions fail. The highest success rates belong to those that operate in good faith. This good faith conduct path exposes any bad faith of the other parties, and, just as importantly, instills in others -- confidence in you.

Because humans are such instinctive liars, we can slip into this deceptive behavior effortlessly. We lie to ourselves about it, making our lies even harder to detect. This creates rationalizations in our C-brains about "why the lies are truths" which in turn creates even more distance from the truth.

Bad faith includes bragging about abilities you don't have, lying about the circumstance, intentions, resources, or known defects in the relationship, business plan, criminal history, etc. Bad faith is a doctor taking money from a drug company and lending his name to research, which he did not perform. Bad faith is being President of the U.S. and taking money from others in order to allow them to influence policy, escape jail, or use the prestige of the office. Bad faith is telling your girlfriend you love her--when you don't. It's changing your golf score. It's holding onto a contract for a few days so you get the commission instead of someone else.

> In the long run, you cannot out-lie, out-manipulate, or out-maneuver other humans no matter how clever you think you are. In the end, each and every person who acts in bad faith pays a price. This is a universal core truth. Even if everybody signs a contract, gets married, etc., how long will it last if the value or love isn't there? If the love or value is there—why lie in the first place?

---

2. Try to manage your behavior, not the outcome.

Fifty thousand self-help books can't be wrong — right? Of course they are wrong. Forget all that nonsense. Nobody can guarantee an outcome of even something simple, much less the incredible range of possibilities available in this world. Goals are good; nobody ever built a skyscraper without setting out to build one. But, remain flexible. Listen to what others are telling you. Evaluate who acts in good faith. Modify your plans during life to find a compromise between your goals, and the goals *others* you trust, think are realistic.

Not a single self-help book tells you this — because they all lie. They all tell you what you want to hear. It sells better. Just managing your own behavior is challenge enough. Most people aren't unhappy because they don't have enough to eat or a roof over their head--in the U.S. anyway. Most people are unhappy because they attempt to manage outcomes they cannot control, and then punish themselves and everybody around them when they don't end up with their desired result.

Anger and unhappiness are most often caused by the denial of expectation.

Consider this point when deciding the importance of acting in good faith. Imagine the complexity of a major airport on a busy day--no human can comprehend managing such a complex thing alone. But if everybody works on a specific task in good faith, managing their part instead of the outcome, things miraculously work well at a very sophisticated level. Most truly complex systems are self-organizing, not master-planned, even in *apparently-planned* facilities like airports. The human body itself is an amazing example of the "self-organized concept."

Man still hasn't harnessed the productivity potential of different engineering principles/approaches based upon this concept. We currently remain committed to solutions that are completely engineered from wingtip to wingtip. This type of engineering is reaching its limitations, and must evolve for the potential success of large projects in the future.

3. Build your life plan around your family, not your career.

The female brain needs a family to be happy. The male brain needs a family and to work productively to feed that family in order to be happy. It's just as easy to be happy as a welder as it is to be happy as a corporate executive because both jobs put bread on the table. Most people would disagree with this, favoring the CEO model. The opposite is usually true, since many corporate executives sacrifice time with their family and end up unhappy. The wife or husband who knows the other spouse feels that family is first above all else, is much happier. In the end, her/his happiness is yours. Happiness needs no job title.

4. Never quit reading.

Only reading expands the C-brain; TV actually puts it to sleep. The brain cannot watch 30 hours of TV a week without getting weak. Watch "Avalon" by Barry Levinson. It's not just about Baltimore and the immigrant experience as the critics think—it's about the devastating effects of TV on the American family. Let's assume you are genetically faster than I am (virtually everybody is). But, I run for a half hour every night and you watch TV and drink beer. When we have a foot race, I will win the race.

Reading is *mental* exercise. If you want to be in the race in a technologically advanced society, you must exercise your C-brain "muscle." Your brain must convert all these words in books into pictures, images, and ideas. That conversion process takes work, work that makes the mind strong.

Your brain is your greatest asset. Don't drown it in alcohol, denial and re-runs. The entertainment powers in Hollywood and New York want to create an endless media experience for you—so they can sell you products advertised in their entertainment experience. Many of you are in denial about how lazy this makes your brain, and so you go along with it. It's a trap so clever I really don't think even this book will make a difference for some of you. This book gives you the truth. What you do with it is up to you.

5. Your children need you.

Forget about using "quality" time with your children to make up for the fact that you don't spend much time with them. Limited time, "quality" or not, is just one of the endless rationalizations people make to justify being away from their children. If you want happiness, spend time with your kids. No one can tell in advance what's going to be quality time and what isn't. The more time you spend with them, the better the chances of experiencing quality time with them.

If you feel that altering your life to accommodate this principle will crush you in some way, financially or socially, don't follow this advice. But, don't say you weren't warned.

If you forget everything else in this book and only follow the advice in the above paragraph and the following one, it will benefit you more than anything else you will do in your lifetime. It also makes this book worth many times its price.

---

6. What you do really isn't important. _Why_ you do it, is.

If your actions are motivated by love, the outcome will manage itself to everyone's benefit. Otherwise, you're just Homer Simpson in the control room of a nuclear power plant. You've become one more moron that thinks he's smart or clever while the radioactive fuel pile is melting its core. Denial isn't just for the weak and stupid: it's everybody's first-choice weapon. We are inherently weak and stupid—but we don't accept this because of denial.

Even humankind's greatest tragedies can be made right, but only through actions motivated by love. Study President Lincoln. He couldn't win the Civil War until he issued the Emancipation Proclamation to abolish slavery. He didn't issue the proclamation to win the war. In fact, Lincoln calculated it would not make a material difference in the war effort. His advisors warned him it would cost him support for the war. Lincoln freed the slaves by proclamation anyway. The war, the slaughter, the death of his son, all the suffering and misery around him, had opened his eyes to God. The love-motivated option that was offered to Lincoln was the final choice he selected. That act did what 600,000 dead, a river of blood, and countless changes of generals couldn't do: it won the war. Additionally, it gave the American people a purpose for their losses, and, upon his assassination, a martyr.

7. Don't believe what you read in the newspapers.

It pains me to write this because I was once a reporter. The current situation, especially in this country, is tragic concerning the state of Journalism. Reporters write because they want to put their ideas in front of the public, not to inform fairly. Editors and publishers hire reporters based upon this same principle. As a result, the entire system now spews more lies than truth.

The public is partially to blame for this. We support these liars, buy their newspapers and watch news broadcasts in order to hear _our own_ point of view repeated back to us. As I've repeatedly urged you throughout this book, look openly at all sides of every issue. Try to see it from the other person's point of view. _Respect balance, not victory or domination._

Beware of extremists from all sides. If they _insist_ they are right, be especially cautious. Unless they are consistently selecting love-motivated choices, they have little chance of being correct.

I've read a thousand articles on global warming and haven't read one yet that I thought was unbiased. **If you really want to understand what's going on with this issue, research Milankovic's cycle.** He was a Russian scientist who chartered the Earth's temperature cycles and matched them perfectly to three terrestrial phenomena that I will not disclose to you.

It is fascinating science that explains much about global warming. It is easily understandable. Read it for yourself. None of the articles I have read on this subject ever mention Milankovic. That is a truly bizarre omission, since *it is Milankovic's* work that should have provided the scientific foundation for a global warming debate. Milankovic is long dead and had no ax to grind when he published. Look him up; you won't be sorry. Lies, denial, political grandstanding, and multiple agendas: people do it all. They do it brazenly, out in the open…and they aren't even ashamed. If you don't put up with it—it will stop.

Book I, Chapter 9

Drugs, Alcohol and Cigarettes, or Your Future

Here is a myth: Some people can safely drink alcohol and use drugs while other people cannot because they can't "control it." I'm not going to get into a long-winded scientific explanation about why that is wrong. Read the statements below and *watch for them in the humans around you*. Eventually, you will come to *know* the truth of the following statements. Or, perhaps, because of denial, you will not.

1. The human brain is affected each and every time you alter its consciousness with drugs or alcohol.

2. The human brain, after enough exposure to blood-borne active neural contaminants, will eventually become dependent or irreversibly changed by that exposure. This change starts with the first drink or smoke, and is cumulative. The alterations that occur are permanent because your "consciousness" is a blend of the physical organ (C/D brain) fed by your blood supply and your life experiences, the metaphysical side, combined. This melding of the physical and the metaphysical creates the filter through which you perceive yourself and your surroundings.

   Your perception of reality, your data-input mechanism itself, is altered by the impairment substance. This may appear obvious and even the point of doing drugs, but the long-term changes in drug users are not apparent from their point of view. Hence, drugs represent a kind of danger from which the brain is not physically constructed to protect you. In fact, the brain's powerful denial mechanisms, a result of the C- & D-brain setup, make it especially vulnerable to this form of attack.

   Other experiences change your mental filter as well. U*sually,* experiential-related change is a positive refinement of the filter through *indirect* influences, rather than direct blood-borne influences. Drugs also alter your mental filter through mood change *while* altering the brain cells physically at the same time. A movie can give you a dose of false reality for entertainment, escape, and enlightenment value, and that is also *accompanied by a chemical experience/reaction in the brain --- just as drugs cause.* In both cases the brains cells are changed by the combination of the physical and emotional experience. The differences begin after the theatre lights brighten at the movie's conclusion.

The normal brain returns to reality and files away (or discards) the memory of the emotional movie experience. The drug user experiences a chemical dosage from the blood-borne neurotransmitter many hundreds or thousands of times more concentrated than the brain's comparable chemical release during a tear jerker of a movie. The worst problem with drugs and/or alcohol is that the brain knows how to repeat the high. In the case of ballet, theatre, and other "soft" entertainment highs, they are not easily or reliably repeated, denying the brain an easy path to addiction.

Not so with the "hard" highs:
    Gambling, (a mad dash of heart-pumping adrenaline)
    Cigarettes (a nicotine jazz)
    Pot (calming, mellow lift)
    Alcohol (loosened C-brain control)

Then there are the Opiates, which literally take away the pain, physical and emotional. But the pain comes back. When you arrive at this stage, you're done unless you get help. The brain in this situation is in over its head. There's more: PCP, amphetamines, cocaine, etc., all killers, literally. Most importantly, judgment-altering drugs kill the brain spiritually because they substitute the drug for a higher power. They in essence become the higher power, blocking our mind's ability to hear God talk to us. This is why 12-step programs need a higher power in order to work.

When brains cells become physically dependent upon a blood-borne chemical, the user loses the will to choose anything else. We are free to choose love; we have free will. If we've already chosen *anything* else, we cannot choose love. Remember; two variables that's all. This actually ties to the Unified Field Theory (later in the book) and the Peace Machine Hypothesis.

Sacrifice, productive work, competition, fear of death: all these important C- & D-brain elements are discarded by a brain that feels chemically elated and/or crashed most of the time. Drug addicts exhibit this mental betrayal as psychotic selfishness. We can see the necessity for C & D balance here. In other words, all joy and no pain is not utopia or love.

3. The obvious intrusions of the real world into a chemically-induced

false reality are covered up by our powerful denial mechanism. The C- and D-brain denial mechanisms, although always on and working, rarely work in concert, as this would greatly endanger the human. Alcohol and drugs are the exception to this rule! Hence, the abuse of these substances unleashes the enormous power of consolidated denial in the addicted brain. Alcoholics that admit to the disease and voluntarily attend meetings are C-brain non-deniers, but are usually still D-brain deniers, which is why relapse is common.

4. For the combined reasons above, *most* people who start down this path die prematurely from it. Put together liver disease, pancreatic disease, auto deaths, suicide, stress-related diseases, cancer, and all the other health issues encountered by the millions of addicted and soon to be addicted, and you have the most massive health epidemic ever conceived—unless we bring back smallpox and polio. Drug and alcohol addiction kills millions *more* each year than war.

5. **Beware of alcohol and drugs during your lifetime, unless of course you simply never start; then it's a breeze. The brain doesn't miss a chemical reaction it never experienced.**

    But what if either I have already started using, or my husband/significant other is already an alcoholic/drug user? Then my advice is simple and based upon some realities you must accept as true. If you are desperate enough, you'll accept the advice below. If not, there's nothing anyone can do until you are ready to make a change. Your denial wall must dissolve before you can concentrate on making yourself healthy.

If *You* Are the Addict:

   I. Don't trust your brain. It will literally kill you. In fact, it is already well on the way to doing that in order to continue getting the impairment substance. *You don't have to be "hungry" all the time.* If you quit your addiction, you can get free. But, it requires abandoning your ego, defenses, and denials -- part of the price you must be willing to pay. Why? So, you can accept a higher power *other* than the drug.

   II. Surrender to a program. Most programs are 12-step type programs, which work to varying degrees. The Scientology drug program is also fine. They use "brain washing," but right now your brain needs a washing. *Twelve-step programs only* work when you place a higher power in charge. The programs are useless without it. Think about the implications of that point. It is a testament to the weakness of the mind and to God's power, His choice of love. It serves as evidence that He exists and is love.

6. Once you're sober, don't trust your brain with mind-altering substances. Not even once. This isn't just a commitment; it's a process you have to evolve through, and it requires enrolling in a program 99% of the time. Sobriety without a higher power is just one C-brain lapse of control away from a D-brain-instigated relapse. A D-brain committed to a higher power becomes the C-brain's ally in the fight against the impairment substance. Only then, does the addict have a chance.

If You Love the Addict:

I. Trust your instincts -- that would be your D-brain talking; and it knows when the addict is lying, high, or dangerous. If "instinct" warns you to leave, or take the kids and leave, do it.

II. Surrender to Al-Alon. You can find this program in the phone book or on the Internet. It will teach your C-brain why your D-brain holds onto the (addicted) person. Then you get to choose: hold on some more or move on. *Neither choice is wrong if motivated by love.*

III. Really start over. I mean this literally. The chances that your D-brain will repeat a similar error are huge. The secret to not making the same mistake is actually simple. Motivate your longing for a new relationship with love, not by what you hope to gain. In this manner, you'll break free of your destructive cycle. Remember, you chose the addict in the first place, and you will choose one again if there is no real change in your life. God is Love and Love is God.

7. Why ruin your life? If you don't already drink or smoke, **don't start**.

8. Marijuana may be the only exception to everything above that I know. I believe mankind has been using it long enough that it actually has

some medicinal benefits, particularly in the vision and nausea relief areas. But I am certainly in denial, which is why I am including the following disclaimer: I smoked pot and none of this information about pot was in the Revelation. This part is just my opinion, and Dr. Richards tells me, it is bad for me. See how clear my thinking was right up to the moment I got to *my* habit and immediately I started rationalizing with my C-brain! Knowing about the C- & D-brain set-up *will not automatically* make you in control of it.

Naturally **I would like to know the truth** about pot, but the U.S. government just won't do a big scientific, unbiased marijuana study. They literally study things that might affect only a few hundred or a few thousand people, but have never studied the overall health effects of a drug used by *tens of millions* of Americans. Some states have even made medical marijuana legal, and still the government will not do a big study. Some states are suing the U.S. government to clarify their position on this plant and still the government will not do a big study: more proof, in my opinion, that denial isn't just in our heads but in our laws, institutions, and policy decisions.

What if some part of cannabis smoking is beneficial, symptom-mitigating (nausea) or helpful to people like cancer patients? What if letting pot get distributed in a manner in which it didn't have to be smoked, saved lives? What if it was the largest cash crop in the U.S. and no tax dollars whatsoever were collected on that portion of the economy, but billions were spent on enforcement and incarceration? What if drug gangs, like the underworld during prohibition, are largely funded by marijuana sales (two-thirds of their revenue comes from pot according to the government), and they use this money to build large sophisticated, murderous organizations that threaten civil order? Lots of government issues/denial here in my opinion.

9. Last but not least on the topic of drugs and alcohol, I leave you with this very sobering thought: 20,000 to 25,000 people, on average, are killed every year in the U.S. in automobile accidents attributable to alcohol. If people really were as worried about the loss of life as we pretend to be, there would be mandated alcohol-lock-out devices on every car sold. You could even charge the convicted drunk drivers, so that the devices would cost nothing extra for most car purchasers. Instead, we moan and wail about the tragedies and keep on drinking and driving. Oh, the denial.

The Revelation

Book I, Chapter 10

Guns and Butter

A healthy human's first D-brain priority is the health and safety of his family. Owning a firearm is one step he/she can take toward this goal. Hunting weapons are an extension of the D-brain meat-gathering priorities, and therefore should also be allowed by law.

A handgun for home defense should be allowed everywhere. This makes the homeowner roughly equal to the stronger, fiercer home invader, despite differences in physical size or age. This is an essential right.

Logically, anyone convicted of a felony, under age 25, having any history of mental illness, or unwilling to take a gun safety course, should be denied this right. Everything beyond weapons appropriate for hunting and self-protection, such as automatic weapons, Teflon coated bullets, bazookas, grenades, etc., is unnecessary for this D-brain primary purpose, and therefore *should be prohibited*. This isn't really a question of whether it's in the Constitution or not. It's a question of getting the balance correct so man's core values, rights, and D-brain behavior patterning closely match the law. I believe the outline above is the surest way to do this, while staying true to our brain structure and evolutionarily determined needs. The Supreme Court ruling in Heller vs. The Government of D.C. was based on exactly these principles—the Founding Fathers envisioned firearms being used as a tool to defend the body, home, family, and country. C-brain-heavy liberals will howl over this section because they are Collectivists. They believe all men and women can get along without guns if we just had good enough economic systems (no poverty), good enough social systems (no war), and good enough education systems (so everybody can be Liberal like them). Unfortunately, they are wrong!

Don't just review the violent-crime numbers. Read the details about *why* people kill each other, and you can see why the C-brain in most liberals, except liberal hunters, is in denial about the presence and motives of our D-brains. Why are liberals in denial about this issue? The reason is simple: if they faced the innate human truths on this topic, a lot of their closely held beliefs would unravel. Washington D.C., aside from being the capital of the U.S., was the murder capital per capita of the U.S. for quite a while.

Liberal politicians, with the influential assistance of *The Washington Post*, essentially colluded to make handguns illegal in D.C. before the Heller ruling. A leading liberal columnist, William Raspberry of *The Washington Post*, championed this cause for years in the Op-ed pages of that newspaper. When his home was invaded, he took out *his illegally owned handgun* and shot at the invaders! His C-brain was passionately writing editorial columns for years saying people should not be allowed to have a gun at home for self-defense while his D-brain spent money to buy a gun, trained him to shoot a gun, and then actually used a gun on a human who threatened him by invading his home. Is Mr. Raspberry a hypocrite? By Conservative standards, he represents the worst form of hypocrite.

But, *like* Mr. Jefferson (a Founding Father who also may have fathered one of Mr. Raspberry's relatives), I don't think he is a hypocrite in the Revelation sense of the word. His C-brain truly believed what he wrote in the *Washington Post*. His D-brain, of which he was blissfully unaware through the usual C-brain denial, was unwilling to be victimized by a potential criminal break-in that might kill him or his family just because the weaker side of his brain, which happened to work for a newspaper, believed the gun control nonsense his C-brain wrote.

Always remember the enormous power of self-lying and self-denial in our evolved *Peace Machine* brain scheme. Conservatives are equally hypocritical. They're just in denial about the other side of those same issues. It is a historical and evolutionary fact: the D-brain will always get its way. If it weren't true, we would not have survived as a species until today. People, who want to kill, will. People, who want to use force to get their way, will. People, who want to take what you own, even if it means possibly killing you, will. There's nothing wrong with giving every man and woman who qualifies and wants protection, the *tool* to keep his or her family from harm -- including Mr. Raspberry. Obviously, we do not need a tool that can kill dozens of people at once or kill Kevlar-covered law enforcement officers, because neither dozens of people nor the police are likely to be a threat to our family or ourselves as ordinary citizens.

Some conservatives led by the National Rifle Association, disagree with reasonable restrictions like those above, and would likely spend money against political candidates who propose the above balanced solution. The right of law-abiding individuals to own weapons for self-defense is inalienable and necessary given human behavior. The right of an individual to own weapons that can quickly kill masses of people in the name of self-defense is a mistake.

BOOK II

Market Fences & Economics

Book II, Chapter 1
Deer Meat, Money

As important as people are in your life, so is money. Money today is our ancestors' "deer meat." Without deer meat (food) we die--right? This is exactly what your D-brain believes. So, for now, we won't contradict it.

The economy not only feeds us, it is the heart of a complex relationship between humans and their environment. This relationship mirrors the same universal truths that we already discussed concerning the human mind. Furthermore, this relationship is mirrored again in what will be revealed to exist in the gas cloud that makes up the universe. Once you understand the repeated connections and the mirroring of these basic relationships, you can predict what the different parties of any relationship will do. Therein lies one of the immense powers in understanding the Revelation.

Evolution has molded the male brain into a hunter and provider. When you do well at work, earn money, promotions and praise, it "feels" to your D-brain exactly as if you have killed lots of animals for meat. The sense of satisfaction men feel from productive work and/or a kill is a neurochemical reward from the D-brain encouraging that particular behavior.

Female D-brains produce a similar reward of emotional satisfaction in association with having and raising children. Women are not as rewarded by the D-brain for hunter/killer work activity. Hence, women are not as driven to compete in a work environment.

This is designed in the female/nurturer/mother's brain for species survival. This also accounts for the lack of deep-seated anger among women about work and pay disparity. Differing D-brain priorities have been evolutionarily determined to ensure that a woman's family takes priority in her brain over pay scales.

If women decide differently, we are doomed as a species, because men are unlikely, also for evolutionary reasons, to pick up the slack. Once we as a society successfully convince the majority of female brains that something else is more important than having and raising children, we've harmed our species much worse than global warming is predicted to harm it. Why is this?

The D-brain (male or female) can be loyal to only one core purpose. The male D-brain "hunter/killer" mind is counter-balanced by its social C-brain Peace Machine. The female D-brain "family-centered mind" is counter-balanced by its "love" or manipulation based C-brain Peace Machine.

The male of our species, historically responsible for hunting and defending the family, has necessarily been counter-balanced by the family-motivated female that nests and produces babies. These two sides make a whole.

Mated Brains are Productive, Healthy, and Likely to Procreate

One major flaw in the way we currently look at human relationships lies in the fact that we falsely believe male and female brains are interchangeable. Instead, each is a part of the human "whole brain unit." You must accept the human brain from this point of observation if you want to understand it completely. The male and female brains, already different on the D and C level, change/evolve further after the birth or adoption of a child.

Men and women have 2 brains each: a C and D. When they successfully mate, those four brains make up one complete working human unit from an evolutionary standpoint. When the male earns, he is a hero to the family and can father extra children. Women become more attracted to him and want to have his children. Why?

As previously explained, men are basically killers. The D-brain program plays like a looped recording in every male's head, trying to keep him alive, mainly by killing most everything around him. That program is interrupted only by the powerful urge to mate. Modern males are a little different. Most of them have strong C-brains, which restrain and partially lessen the daily influence of D-brain thinking. Some men are further restrained by the influence of love from women-of-significance to them: wife, mother, or daughter.

Some professional male basketball players, the epitome of modern strength, aggression, wealth, and fame, claim to have had sex with over 10,000 different women. Though many media pundits question these numbers, I believe them. This is not a freakish number that makes no sense. Rather, it is evidence that the D-brain is alive and well in the world we live in. I doubt any famous female athlete will ever honestly admit to such a feat with male partners—because her brain doesn't work that way. Mating alone is not her goal. Having children with a successful male is her goal.

Because of the tradeoffs between the sexes, men have to create "wealth" to be happy. That requires more than ancient hunting skills in our current tool-heavy world. This can be very frustrating to D-brain dominated males that don't "fit in" when they can't hunt or kill their way to the top. These males usually end up in the incarceration cycle: go into jail, come out, go back in, and repeat.

By its nature, this discussion must also be political. Politics interfaces with money on almost every level. It is important to remember that this book doesn't endorse or support any existing political, economic or philosophic ideology. None of them incorporate the C-brain and D-brain explanation, and therefore their current theories cannot be fully accurate or correct in their representations.

Politics is the struggle of ideology: Communism vs. Capitalism and so forth. In the *Peace Machine* understanding, things are not that complicated. If you don't believe this, read just one of Lenin's books on revolution. Not even one percent of what he wrote is relevant today; in fact, most of it doesn't even make sense anymore. The only struggle is between the C-brain and D-brain. Since we already know what each brain really wants, we have gained deep insight into political cause and effect. It will not be hard for you to extrapolate from this setup how large groups of humans will react or behave. These dynamics become clear as you begin viewing the world through a C & D-brain perspective.

Humans survived in small tribal groups for a long time under a "Might was right" rule of order. Leadership was seized, not finessed from voter groups. This is tyranny, the original form of government. It was understood and accepted because it mimicked the original form of order: the family structure.

Humans inherit in our D-brain programming a sense of comfort when living in a societal environment with strong leadership because it is just like the primitive family. Politics today, still reflect this reality.

For a really, really long time, the D-brain was so dominant in our daily activities that tyranny was the only form of government. It wore many names--Feudalism, Monarchy, Communism, Fascism--all were just different forms of tyranny. The only difference was the method of *selecting* a tyrant. In

Monarchies, the primitive family model literally was adopted as a form of government; bloodline was the selection method. In a Monarchy, "The King is the Law" was actual policy of the government. In Communism, it is the politburo that selects. But in each case, the end product was the same: a tyrant who made decisions for people with little if any actual power-sharing between the tyrant and the subjects. D-brain heavy men favor this form of government: strong, might is right, Man/Dad at top, religious, conservative.

With the advanced development and exercise of the C-brain, mostly through education and reading, the idea gained strength that men could choose their own leaders and enact their own laws. This is a very important milestone, marking the first point in time that the societal C-brain declared it was in charge and no longer just a tool of the D-brain. It doesn't matter that this was a lie. While the C-brain never completely rules the D-brain, most C-brain heavy people believe it does.

Wall Street panics demonstrate this relationship perfectly. People use every C-brain calculation from P/E ratios to technically graphing price floor supports in order to justify investing more money while the market is climbing. But when fear hits, for any reason, people sell irrationally.

Fear, absent training, puts the D-brain in charge, and it does what in a primitive or highly competitive environment is the smart bet: cut your losses. Dump. Sell. The D-brain is literally in charge when you call your broker in a cold sweat and tell him "Get me out." The C-brain literally has little or only limited memory of our behavior in these moments, we self-deny constantly on the existence of these memory issues. Despite these kinds of problems, the C- and D-brain split has helped human comfort and society more than any other adaptation, ever.

The C-brain's delusional decision that it is in charge was more of a positive development for the C-brain than it was a loss for the D-brain *because the C-brain can do something the D-brain cannot: it can read and write.* This ability turned out to make humans better hunters than the D-brain could ever become on its own. A vicious man with a rock is no match for a gentlemen hunter with a gun. This forced the D-brain to give up some dominance and to allow for internal power sharing among people who used their C-brains to acquire food, fuel, and goods.

Democracy is a societal mirroring of our internal form of power sharing we evolved as humans: the D- vs. C-brain. C-brain dominant people living in a democracy do not fear their representative government as much as they fear tyranny. Conversely, the ancient D-brain did not fear a tyrannical state. In fact, it preferred tyranny because it naturally understood this model, as it resembled how it managed its own family. The more developed the C-brain became, the more it feared a state that can exert its will over a man and his family without just cause--i.e. C-brain-agreed-upon laws. Laws and letters, as well as ideas of fairness, are C-brain constructions. As we will see, they often lead to bizarre results from a D-brain point of view.

When people achieve high literacy rates and develop strong C-brains, the newly empowered C-brain has a desire to escape the fear and intellectual claustrophobia of a dictatorship, and to live in a land where the family is free from tyranny. This goal, especially when involving children, can get D-brain support.

This creates men like George Washington, whose C-brain was "…the father of all Democracy." in the words of Lafayette, yet, whose D-brain, like Jefferson's, still owned slaves *to the day he died*. Don't misunderstand me: George Washington is one of the greatest men that ever lived. This is an opinion I would be hard pressed to defend, if I didn't understand the *Peace Machine*. In reality, we only have two government models:

Model 1	**Model 2**
Tyranny	**Power-Sharing**
Monarchy	Magna Carta model
Fascism	Parliamentary
Communism	Representative Democracy

The models on the left evolve into models on the right, usually through civil war or revolutionary war. Tyrants generally chose economic systems like communism that do not encourage competitive behavior and are therefore less productive. Because of this diminished productivity, tyrants tend to have problems feeding their people.

This makes them unpopular with both the C- and D-brains of their own populations, which they then must control with force, fear, and incarceration: jail, Gulag, Bastille, etc.

China is an important exception to this rule. The ruling Chinese have chosen to use communism only as a political tool. They continue to control their people through fear and intimidation while using capitalism, usually a tool of democracies, to drive up productivity and national wealth.

Putting wealth in the hands of a D-brain, leadership-driven system will produce great success in the short term. If the Chinese people feel this wealth is percolating down to them, their D-brains become/remain happy, and they will not mind the fact that they do not get to pick their own leaders or laws.

Why? Because "The D-brain always wins."

The Tiananmen Square protests began as an outpouring of grief—over the death of long-time communist leader, Hu Yaobang. It ended with tanks rolling over unarmed protestors. The Communist Party in China still does not understand how an outpouring of grief over the death of a communist leader morphed into a democracy protest, so they are no longer taking any chances and simply tamp down *all protest aggressively.*

During the 2008 Olympics in China, two elderly women, one 79 years old, and the other 77, were sentenced to a year of labor, because they applied for a permit--to protest in a Chinese government-officially-sanctioned protest area. The Olympic Organizing Committee only gave China the right to host the Olympics under the condition that people would be allowed to protest and demonstrate.

In practice, the Chinese government simply jailed, exported, or silenced every person applying for a protest permit, except the Western journalist who applied for one. He didn't get it either—but he wasn't jailed.

The D-brain political system employed by China today is a tyranny. The D-brain model likes a cave full of increasing amounts of "deer meat" and clearly understands a strong leadership model. Communism is certainly all that, and more. As long as the government is making the people wealthier each year, China's leaders/fathers/tyrants will remain in power. But, every economy eventually stumbles, and when they do...

By numbers, the Chinese people are among the worlds most literate, and they are *becoming among its most productive.* This C-brain typhoon is strengthening, while waiting for its moment in history. What will occur in China will change the course of history for the human race and the world. There is no doubt that the Chinese populace's C-brains will eventually assert themselves, demanding some type of power-sharing with their government.

The big question is when and how the Chinese government will respond? Will it act as Russian President Mikhail Gorbachev did in Communist Russia, by surrendering power peacefully to the people? Or, will it take a civil war to force power-sharing?

To quote Winston Churchill, "Indeed, it has been said that democracy is the worst form of Government except all those other forms that have been tried from time to time." While Democracy does not always bring about good leadership, it does bring about consequences for which the people share responsibility. This leads to good and bad leaders, but usually not murderous or war-mongering leaders.

Most of the world's populations currently don't live in nations where they share power with their leaders. In these nations, the people are not responsible for the country's actions. But in the end, they are usually the victims of their leaders' decisions. What does this discussion of C- and D-brains and forms of government have to do with deer meat or economics?

Most societies make the transition from the tyrant stage when enough of their citizens graduate from being primarily poor, uneducated, and self-sufficient Basic Human Units or **BHU's** (D-brain heavy) into more highly-developed C-brain-driven Competitive Human Units or **CHU's**. There is much more on this in the next chapters.

Young democracies usually experience a cluster of wars that break out in the first 50 years of their existence, because other neighboring nations' tyrants hope to de-stabilize or conquer them—before the power-sharing idea spreads to their own populations.

Here comes India

With so much written about China in this book, and with my desire to limit the book's size, I have given very little space to the country of India. I will correct this oversight for the following reason: India is the sleeping giant of this century. She sits in a unique position, very similar to America in the last century. India's religious split is her Achilles heel. Watch for India in the coming years.

1. India already has a social system better designed to minimize D- and C-brain conflicts compared to other industrialized or partially Muslim countries. India has the about the same percentage of Muslims in a primarily Hindu country as the U.S. has blacks among a primarily white population. When it comes to creating the family structure, Indians don't fool around with self-selecting D-brain infatuations. The parents arrange the marriages of their children in a way expected to benefit the entire family of both groom and bride. This is brilliant from a C/D-brain viewpoint because the parents look logically (C-brain) for compatibility in social-economic status, religion, and family stature. They count on the D-brain sex mechanisms of the

bride and groom to fuse together two people who may have never met before the wedding. The D-brain is very good at this type of fusion *because it was designed to do it.* This arrangement takes into account both C- and D-brain needs, and is why India's long-term marital success rate is *double* the U.S. rate. Because of the C/D brain relationship, we know that family units are the structure upon which society is built. Having marriage customs that put C/D synergy into practice is highly productive. When the percentage of self-selected marriages in India rises, so will the divorce rate.

2. India elects her leaders. The growing C-brain population is working productively and becoming educated in greater numbers while embracing productivity improvements systemically. They will become a huge race of highly productive, highly educated, highly motivated, God fearing, God loving, and God understanding people. All of these elements, including India's religious minority, will make her a bigger, though less wealthy, fast growing version of the U.S. on the Asian continent. The American people already have an ally in this race/country. We need to cement those ties in the coming years to ensure our own survival. As we struggle to modernize but not lose our way culturally in the U.S., so will this parallel struggle be ongoing in India.

Now that we've thrown out those unclear and unneeded differentiations of government philosophies and replaced them with just two: tyranny and shared power, we can do the same in the next chapter for the myriad of theoretically different economic systems. There are also only two.

Book II, Chapter 2

Collectivism vs. Competition/Productivity

There are two main disciplines of economic thought today. The first, Collectivism, is the most popular. Of the two, it is the most fatally flawed. Collectivists are people whose economic outlook is inspired by their C-brain. They tend to be intellectual, liberal, and communal. They are not killers, but they get a lot of people killed. In some countries, they are in charge by virtue of a dictatorship (Castro). In other countries like Italy, the communists, collectivists, and socialists come in and out of power on a constant basis.

Fair vs. Successful

Remember that evolutionary behavior, regardless of our "politically correct" opinion of it, is always programmed toward the successful. The programming itself is the result of very long-term evolutionary "Fundamentals" that "push" adaptations out in front of them. Fundamentals are the most basic of biologic evolution. They apply across many species and represent the aggregate of the biologic evolutionary response at any one time in history. Presently, the "big and strong" Fundamental is out. It was wiped out by a meteor strike 65 million years ago. To picture the Revelation image of a Fundamental, imagine water gently flowing past the blunt bow on a slow-moving boat. The distorted water would be the adaptations. The bow of the boat would be the Fundamental. One such Fundamental used by the dinosaurs was a combination of strength and size. In the predators of these species, it was combined with speed, ferocity, biting power, and killer instinct.

The Fundamental beneath man's ascension to today's advanced consciousness is advanced nurturing and/or love. We should carefully consider the implications of this from two important perspectives. The first is that we share DNA-inherited memory with those dinosaurs, in our D-brain. Secondly, the difference in the way these two Fundamentals operate is mirrored in our C- & D-brain structure. All advanced-consciousness creatures are guided by behavior that was successful in their ancestors, behaviors that are therefore integral and modeled into their present-day programming.

Evolution has no boundaries; this applies to plants as well as animals. It is evolution that stretches a black hole into a multi-field universe (matter and energy are two of these fields). Biologic evolution takes matter-based life through changes without regard for "fair." This is hard for a C-brain dominant person to understand. You need to remind your brain constantly that "fair" and "successful" are not the same thing. "Successful" is the identifying tag worn by the surviving members of every competition. It is the end result of every natural or artificially-created process and competition.

"Fair" is an idea, formed by people in a competition about what the results of the competition *should* be. "Successful" is a measurement, even if only a relative one, of something real. "Fair" is just a perception, a C-brain illusion.

It is mathematically possible to learn from a measurement, not an opinion. "Fair/Unfair" is only opinion about the result of a contest *formed by viewers and some members of the contest*. If the leaders of your County Council, Union, Country, State, or PTA are calling for "Fair," beware. When "Fair/Unfair" is given control over a system's direction, the system itself will spiral down, as a mathematical certainty, to lower productivity and certain breakdown. Put another way: when income distribution in a "Fair" system is nearly equal, the biggest argument won't be over how to increase total GDP through marginal increases in productivity, but rather over how to distribute the smallest remaining differences more evenly. Put "Successful" in charge, and everybody gets an improved standard of living.

"Successful" equals increased marginal productivity. Since competition, with adequate anti-trust protections provides the environment required for the greatest productivity improvements, it creates wealth the fastest. The Chinese are good Communists and good people: they gave Collectivism a solid try for 30 years under Mao, and it didn't create widespread wealth. In fact, it was a struggle to feed the population. The very same people then tried capitalism, competition, and productivity. The result of that experimentation was the creation of more wealth, faster, and for more total people, than has ever been done by humans.

Theoretical communists such as Marx, Lenin, and Trotsky were collectivists. Since tyrants eventually rule real-life communist societies, 2nd generation (and later) communist leaders tend to be very D-brain. In Stalin, the Russian people got the worst possible outcome: a D-brain individual using violence and murder to implement a C-brain scheme of collectivizing farms that resulted in the death of millions who resisted this "land reform." Millions of Russian survivors subsequently starved to death because of how marginal productivity suffered as the result of this C-brain collectivist scheme. This ugly combination of C-brain scheming and D-brain tyrant leaders can kill millions; and, in Russia, it did. As an aside, Stalin sent an assassin to kill Trotsky, who was exiled in Mexico, while Stalin simultaneously implemented Trotsky's plan to collectivize the farms. This represents a C-brain complement paired with a D-brain motivated assassination.

Since the collectivist philosophy is based in the C-brain, and the D-brain controls human behavior, collectivists are great at designing solutions that can't work in the real world. Some huge examples include: The Missouri Compromise, AFDC guidelines in the U.S., and just about everything in Russia between 1917 and 1985. Why does the C-brain design such flawed plans that actually threaten survival? They start as plans to promote "Fair" and "Deer Meat" for everyone. Because C-brain solutions work well for certain activities like tool making, reading, mathematics, etc, it is assumed by the C-brain that it can design a solution to any human problem. Pay attention to this principle from now on. Use your C-brain and D-brain appropriately so they can benefit you.

C-brain thinking works well for scientific and technical problems that can be solved by logical thought, the design of a farm tractor for example. Unfortunately, political ideas must be implemented among D-brain influenced humans who do not behave as the C-brain predicts they will. In many cases, it is the manner of implementation of an idea that is the key to its success or failure, not necessarily the weight of the idea itself. Hence, the productivity lost by slaughtering the peasants who worked the land in Russia in order to collectivize the farms far outweighed any benefit that could have resulted from changing the nature of the farming system.

Which C-brain ideas have poor success rates? C-brain designs involving natural systems almost always fail. In these cases, the C-brain scheme is competing with a natural designer (evolution) that has more experience, hence the high failure rate. But, with care, man could overcome this propensity for failure. To succeed, he must conduct his planning and science in a different manner, especially in the area of geo-engineering. Geo-engineering is a science mankind will need in order to survive the next 100 years.

People that can design airplanes should know not to trample each other when exiting a crowded theater when someone yells fire. But, people trample each other to death in such situations most of the time. Why? The D-brain is the one getting you out of that theater.

Conservatives, according to Collectivists, are mean people with horrible ideas about the greedy nature of man. Since the D-brain more closely resembles this description, many conservative political positions, while they appear uncaring to the C-brain, actually work. Since Conservatives don't understand why this is so, they don't have any idea which Conservative principles to champion and which Liberal ones would work better if tried.

As a result, the Conservatives cling to their combination of successful and unsuccessful principles and policies and criticize the Liberals. The Liberals, meanwhile, cling to their combination of successful and unsuccessful principles and policies and name-call the Conservatives. *The most important example of this disconnect between philosophical outlook and political practice is the productivity principle.* It is motivated in the D-brain by the desire to hunt, to collect deer meat, to accomplish something of concrete value/work each day.

Conservatives are ostensibly the standard-bearers for the principle of productivity, namely that improved productivity will benefit all. Although they pretend to understand this idea, it's just lip service because they don't understand what an important concept they have adopted. As a result, they don't do it justice with their policies. In fact, in the pantheon of ideas that define them, other less important principles of the conservative movement occupy more important positions than does productivity. This makes them lousy champions of the "Productivity Principle" philosophy, which truly represents wealth for all. In the end, conservatives, usually Republican, have to choose between big business, which by its very nature is anti-competitive and wasteful of resources, or the Productivity Principle. They continually disappoint and toss aside productivity in favor of political donations and free rides on corporate aircraft, to places that just happen to have wonderful golf courses.

In a Productivity Principle-driven universe, the one we live in, only robust competition for the re-use of poorly-utilized or excess resources can produce marginal gains in wealth.

This follows because nature (read evolution) "loathes" under-utilized resources. Too much of anything that is not being consumed leads to contamination of the life cycle for everything around it. A lack of competition leads to stagnation and eventually death for many creatures, regardless of the proximate cause. Competition is the most efficient way to force all the organisms present in the environment to use the available resources completely. Man had no serious competitors once he transitioned into a super predator after 1422 and the invention of the printing press. Widespread reading and writing were not possible before books and reading materials could be manufactured. This led to the widespread use of tools. Hence, man was finally able to abuse the resources available, and he did, killing thousands of other species.

Book II, Chapter 3
Combined Laws Of Thermodynamics or CLOT

The combined laws of thermodynamics (CLOT) are of great importance because they underlie the key principles in economics and physics. They form the foundation of the Unified Field Theory and prove the impossibility of time travel. Rather than paraphrase them, I will quote from a paper written by Mr. Max Planck, one of the greatest minds that ever lived. I am shortening his paper considerably (my apologies to Mr. Planck) in order to keep this chapter from becoming a physics lecture.

Point 1. "An achievement which is closely associated with the discovery of the principle of energy, and which is one of the most important for the theory of heat, is the proposition that the total energy of a gas depends only on the temperature, and not on the volume."

Point 2. "It is in no way possible to completely reverse any process in which a gas expands without performing work or absorbing heat, i.e. with constant total energy."

Point 3. "The proposition of the two preceding paragraphs, therefore, declare that the generation of heat by friction, the expansion of a gas without the performance of external work and the absorption of heat, the conduction of heat, etc., are irreversible processes."

Point 4. "The second law of thermodynamics states that there exists in nature for each system of bodies a quantity, which by all changes of the system either remains constant (in reversible processes) or increases in value (in irreversible processes). This quantity is called, following Clausius, the entropy of the system....
Since there exists in nature no process entirely free from friction or heat-conduction, all processes which actually take place in nature, if the second law be correct, are in reality irreversible."
Simplified: the combined laws of thermodynamics (CLOT) as applied to

economics describe a gas cloud (our economy) that converts all the energy (productivity) imparted by the cloud particles (e.g. employed/productive humans) to heat (or work output - which equals wealth). This heat created eventually affects the heating of every other cloud particle. This process is unavoidable and irreversible. This has profound applications in economics where work times productivity equals the heat in the cloud that represents total human wealth.

Economies behave as big gas clouds with "heat" equal to "wealth," which is not necessarily the same as money. The human economy, once above a certain minimal population and productivity level, acts in a direct and predictable way to the addition or subtraction of energy (human work or productivity) from the gas cloud (economy). More economic activity (more heat/energy) leads to more production (cloud heat and expansion), resulting in a growing economy. Less economic activity (less heat) leads to decreased output (cooling and shrinking gas cloud/economy) and recession/depression.

Just as importantly, "economic clouds" react in an "ideal" manner when the gas molecules are concentrated enough to constitute a gas cloud, but not so concentrated as to be under high pressure. The following is from a chemistry textbook. Take a general look at these principles: they are very important in our economy. You do not need to understand the specific numbers; the concept described here is what's important.

> Ideal Gas Law: Pressure x Volume = Moles x Ideal Gas Constant x Temperature
>
> Substituting in variables, the formula is: $PV=nRT$
>
> $PV=nRT$. P is pressure. Pressure can be in either atmospheres (atm) or kilopascals (kPa).
>
> V is volume in liters (L). n is the number of moles of the gas.
>
> R is the Ideal Gas Constant. Depending on whether atmosphers or kilospascals were used, the value is either 0.0821 L-atm/mol-K or 8.31 L-kPa/mol-K, respectively.
>
> Temperature is in absolute degrees Kelvin.
> An interesting aspect of the Ideal Gas Law is its flexibility. Its structure contains elements that allow you to solve for other quantities, such as density or

molecular mass.

All this skirts the concept of: What exactly is an ideal gas? An ideal gas is one that exactly conforms to the kinetic theory, as stated by Rudolf Clausius in 1857, which has five key points, and they are:

1. Gases are made of molecules in constant, random movement.

2. The large portion of the volume of a gas is empty space. The volume of all gas molecules, in comparison, is negligible.

3. The molecules show no forces of attraction or repulsion. (The Revelation deviates here.)

4. No energy is lost in collision of molecules; the impacts are completely elastic.

5. The temperature of a gas is the average kinetic energy of all of the molecules.

Non-Ideal Behavior: The Kinetic Theory makes several assumptions about an ideal gas. These cause problems because real gases are not ideal. The main causes of error are related to pressure and temperature.

Pressure. At high-pressures the behavior of real gases changes dramatically from that predicted by the Ideal Gas Law. Under 10 atmospheres of pressure or less, Ideal Gas Law predictions are very close to real amounts and do not generate serious error.

Temperature. When the temperature of a gas is close to its liquefaction point, the behavior is very different from Ideal Gas Law predictions. With increasing temperatures, the Ideal Gas Law predictions become close to real values.

Why? The answer is simple: ideal gases have molecular volume and show no attraction between molecules at any distance; real gas molecules have volume and show attraction at short distances. Let us first consider what pressure does. Pressure at high degrees will bring the molecules very close together. This causes more collisions and also allows the weak attractive forces to come into play. With low temperatures, the molecules do not have enough energy to continue on their path to avoid that attraction.

The first thing to point out at this juncture is that the Revelation shows another connection between these molecules, a metaphysical one, that exists between these gas molecules through the quantum mirroring present in the Singularity, and this force is not accounted for at all by the description above but has been observed by numerous laboratory experiments. You the reader may be asking: Why is this gas science necessary and important? Because this "ideal" concentration range is the only circumstance in which the gas molecules (humans, assets, corporations, etc.) display the maximum amount of distinguish-ability while preserving the efficiency of close contact. Humans need this economic distinguish-ability to identify and suppress less marginally productive behavior that lowers cloud energy and simultaneously leads to behavior that ends with either a bubble or a crash. You can identify when a market is unhealthy, because you will see the distinguish-ability disappear. In a credit market, you would describe this as the absence of a healthy interest spread between the perceived good and bad risks. Indeed, when this occurs, *as it did for years* prior to 2008 with an inverse curve yield, trouble in the economy is sure to follow.

The Basics...

We will call what an average person uses to survive for one year (including the person themselves) a BHU, or annual Basic Human Unit. When an individual produces more than 3 times the amount of a BHU, they become a CHU, or Competitive Human Unit. CHUs have a significantly more powerful positive effect upon their economy (gas cloud) than do BHUs.

To express The Total Wealth (TW) of an economy mathematically: TW equals:

Basic Human Unit production (BHU) *times* **the averaged environmentally based multiplier (Ex) per BHU in that population** *times* **the total number of BHU's** *Plus* **the Capitalist Human Unit production (CHU)** *times* **the considerably higher education based productivity multiplier (Px) per**

The Revelation

CHU in that population) *times* the total number of CHUs.
Therefore, an economy's Total Wealth is described as:
TW = (number of BHUs times Ex) *plus* (number of CHUs times Px)

The CHU portion of the equation is usually vastly bigger than the BHU portion because of the productivity multiplier afforded by education, physical opportunity, and access to investment capital. The product of the CHU portion of the equation is what we call GDP, or the monetized portion of the economy. This equation has vast implications for immigration and education policy.

It is crucially important to understand that you can more easily add to the marginal wealth of a CHU (and most BHUs) **without taking away** from other CHUs or BHUs than you can by attempting to do it through any other method, including transferring the wealth through charity, taxation and re-distribution, war, or theft.

All an individual has to do to increase his/her wealth is to work harder, longer, or more productively. In this manner, each individual adds heat (wealth, deer meat, per capita GDP, whatever you call it) to the total wealth cloud shared by all. Many people mistakenly see economics as a net sum game—e.g. "I'm rich because they are poor." These are "Liberals." The poor Marxist peasant would say "I'm poor because they are rich." Both are wrong. Unfortunately, this fundamental misunderstanding has caused much war and misery.

A million years of small populations struggling to survive in a subsistence environment has made the incorrect belief in "net sum logic" so imprinted in the D-brain, only overwhelming evidence to the contrary can satisfy our C-brain suspicions. This D-brain misconception is why Liberals feel guilty about the material success of their industrialized countries. On the other side, nationalistic leaders capitalize on the fear that their native populations lose when outsiders gain wealth or jobs in order to stir up xenophobia and protectionist sentiment. Both responses to the idea that wealth is a net sum game are painfully destructive to the economy and are based on a false belief. Lord John Maynard Keynes had a difficult time convincing Franklin Roosevelt that the multiplier effect was real and that he could count on government spending to boost GDP by a multiplier that is derivative of marginal propensity to consume. In fact, Roosevelt mentioned to his secretary after the meeting that Keynes didn't make much sense: "…he just talked about a bunch of numbers…" Interestingly, Keynes was not that impressed by Roosevelt either. It's very hard for one nation (while free of self-decline) to lower the marginal productivity of another, *except by war*. Once people are prosperous enough to consume/produce two times their subsistence level of goods and services in one year, a single BHU can become a double BHU.

The guilt, fear, and genetically-inherited hatred we feel for those outside our economic system seeking to increase their wealth is misplaced. In fact, **only** by increasing the marginal productivity of all humans can we add to the world's total wealth. Similarly, decreasing the marginal productivity of any population will subtract from everyone's wealth.

This cannot be done "for" them. The only way out of poverty or subsistence living is for a BHU to self-advance to a double BHU on the way to becoming a CHU. You cannot cheat this system and fool nature/evolution by just giving a BHU money. Money derived from natural resources like oil and a progressive taxation system can be re-distributed among BHUs. But when the oil money -- or tax revenue source -- runs dry: a large percentage of the BHUs living like CHUs will return to poverty. Only by learning to be competitive and therefore more productive through education, opportunity, and some capital, can BHUs *evolve* into CHUs.

Nations that re-distribute wealth in an attempt to buy their way out of this reality constantly undermine productivity, which in turn erodes wealth creation, leading to inevitable economic failure.

When the distribution of resource wealth (e.g., petrol dollars) is concentrated by a nation-state, it appears to contradict this principle. But in reality, since every economy is governed by CLOT, the previously described economic principle is immutable. Virtually all Marxist states (e.g. Cuba, Venezuela, and Korea), most monarchies, and most socialist states (e.g. Mexico), make this mistake. For example, after nationalizing the oil companies into Pemex, Mexico diverted oil income to the state coffers for re-distribution. The oil fields, production and even administration of the company and her assets have consistently declined ever since, with serious drops in productivity/wealth. This type of social/economic engineering lacks a competitive element and will always end in failure. The laws of mathematics and CLOT dictate this.

A BHU cannot grow their personal wealth/work product outside their area of knowledge and experience. The rice farmer growing only enough food for his family must first learn to grow twice as much rice before he can increase production to over 3 times the amount needed. Only after he reaches that 3-

time level of productivity can the BHU advance economically *and through C-brain maturation* to become a CHU. Producing goods or food at this level is the learning experience needed to allow his brain to be a trusted "tool" that can take him to 10 times that production level. BUT, not before! Why?

The rice farmer has a BHU brain that must learn to trust the first competitive/productivity step he has already taken before he can/will take the second step to increase production further. Lessons not learned through experience, immersion, and/or competition don't mold the human brain. This step-wise progression is a D-brain limitation on the C-brain's abilities.

Evolution chose this limitation as a safety mechanism. It would be dangerous for survival if the D-brain relinquished control of "deer meat" -acquiring activities to a brain element, the C-brain, which had not demonstrated convincingly that it could supply enough food and fuel for survival. Micro-banking is based on this step principle…and is why it works well. The other principle that differentiates/distinguishes a successful micro-banking operation from one that fails is women. Women make, keep, and stock the nest. They are responsible, usually loving in their behavior, and are the first to take out small loans in third world countries for productive necessities like sewing machines. And, women pay back the loans.

A BHU will only voluntarily work his or her way out of poverty through the one method that is currently supplying the BHU with food. Most advanced countries' diplomats simply do not understand this. Their attempts to help less-developed countries almost always fail through this lack of understanding.

The world needs BHUs to become CHUs in a specific progression, or else they are not self sustaining in the long run. This is a reflection of brain function and structure, and thus is an immutable economic reality. Furthermore, whether or not the BHUs are in a political power-sharing arrangement during this evolution is largely immaterial.

The last statement above contradicts much current Western democratic thought and writing. But, the poor track record of democratic success in populations unprepared for it (over the past hundred years) clearly illustrates that they are wrong. When the double BHU human moves up to produce an individual GDP of three or more BHUs annually, he/she becomes a Competitive Human Unit, or CHU. These CHUs can use tools/capital and productivity to multiply their efforts further. In the United States, 5% of the world's humans are responsible for 25% of the world's GDP. There are many U.S. CHUs with large productivity multipliers. Some highly productive large corporations build upon their CHU employees to achieve *average* productivity multipliers of 50 times worldwide average GDP. Much political writing and policy today is based upon the assumption that wealth is somehow transferred from more deserving areas of the world into the U.S. where it is gluttonously consumed. This is a ridiculous idea, and virtually all economic data specifically contradict it.

Every time the U.S. has fallen into recession or depression (decreased consumption *and* production), its GDP losses *spread* to the rest of the world, badly damaging other economies.

If economies were truly a net sum game, as the majority of humans falsely believe, then economic losses in the U.S. economy would be offset by gains elsewhere in the world. That has *never* happened! This is also why the current balance of payments/trade deficit explanation as expressed by modern economic theory makes no sense. Countries that suffer from negative trade balances year in and year out seem to have *not* been negatively affected -- and even did well while running large trade imbalances.

If the trade imbalance is caused by the importation of goods or services that increase the purchasers' productivity, the resulting economic warmth in the cloud from increased productivity more than makes up for the cooling the cloud must absorb from the trade deficit. The reason? Productivity gains are cumulative and self-perpetuating, while the loss due to the negative trade balance is a one-time event.

By way of example: If the wheat you sold (exported) in one particular year only fetched $10 and the machine you bought (imported) to improve production cost $20, you would have a negative trade balance of $10. But let's assume the machine was going to net you $5 dollars extra a year over the next ten years through increased productivity. Clearly, the trade deficit was a good thing for the CHU *and* the eventual, total GDP. If a trade imbalance is caused by the importation of goods or services that result in marginally less productive behavior (Champagne) or serves only for the satisfaction of consumption by debt (also Champagne), then the lack of economic gas cloud heating (no increased productivity from the acquisition), in addition to the one time cooling associated with the trade deficit, *can result in total economic cooling*. Because of the *rarity* of this second situation compared to the first example, most economies grow more in the years when they trade, due to productivity gains, and grow less in the years when trade is restricted.

A simple question: Do you actually suffer economically as a result of a trade deficit? The answer, 99% of the time, is no. In fact, countries that trade invariably get much wealthier, than those that don't. If current economic thinking was correct, countries that had trade surpluses would be the wealthiest, non-trading countries the next wealthiest, and those with trade deficits would be the poorest. *In reality, countries that trade a lot are always among the wealthiest; and those that don't trade much are always among the world's poorest.*

This core fact favoring trade has fueled more globalization in the last 20 years than ever before. Trade has created more international wealth than ever before. But, this has also produced excesses and unsustainable habits/behaviors; the most significant of these is the industrialized nations' dependency on oil-producing nations for fuel. This is an example of an irrational and dangerous trade policy that is more likely to cause war than keep people wealthy, warm, fed, and moving.

In many countries, this poorly thought-out dependency on oil has created gigantic trade deficits, which *could become* a threat to economic stability from their sheer size. Irrespective of whether a country has a trade deficit, when a country's population consumes more than it produces, it is going into debt. This total debt is different than a trade deficit. This national debt is exacerbated if the imported item, like oil, does the same amount of work regardless of price.

Investments in imports typically have productivity tied proportionally to price. Higher-priced items would create more short-term debt but generate more long-term productivity. Unfortunately, it is not so with oil. The same gallon of oil releases the same amount of energy whether it costs one dollar a gallon or three. The change upward in the price of oil is a net loss in wealth, that's why oil price spikes "cool" the economy, and low prices "heat" it. Conversely, when one nation's population segment produces more than it consumes, it is collectively saving whether or not the individuals in the group perceive they are actually saving, China during the '90's is a good example. These savings are the source of lending and capital investment for future productivity improvements at home and abroad, and are therefore *critical* to continued wealth formation. This is obvious from even a cursory examination of economic data/history--and yet protectionism continues to spring up as a viable policy option every year around the world. Why? This is the result of D-brain fear—mostly about jobs—and is grossly irrational since *productive* trade is what pushes overall productivity. Trade is one of the greatest wealth builders for humans because it strenuously encourages advancements in productivity. Labor Unions do not understand this essential concept. Their lack of understanding is extremely unfortunate because labor unions have the potential to be one of the great forces for wealth creation on the planet. Instead, the current "big" union anti-free trade stance has paradoxically made them a force *against* wealth and job creation, ultimately leading to the extinction of their own members' jobs and the union's viability.

I want to offer some unabashed and lavish praise on one of the great economic minds alive today, Robert J. Samuelson. Mr. Samuelson isn't just a great economist and thinker, he's an honest man, something I think is rare. I've included for your enjoyment, parts of an article he wrote in 2008 about poverty. It's interesting because Mr. Samuelson has reached the right conclusions about poverty and the importance of trade without learning the *Peace Machine* overlay. Consequently, he admits he does not have a solution to these problems; still, that does not detract from his message. In fact, his analysis clearly highlights the importance of the Revelation's core principles.

Rx for Global Poverty
By Robert J. Samuelson
Wednesday, May 28, 2008

What's the world's greatest moral challenge, as judged by its capacity to inflict human tragedy? It is not, I think, global warming, whose effects -- if they become as grim as predicted -- will occur over many years and provide societies time to adapt. A case can be made for preventing nuclear proliferation, which threatens untold deaths and a collapse of the world economy. But the most urgent present moral challenge, I submit, is the most obvious: global poverty.

There are roughly 6 billion people on the planet; in 2004, perhaps 2.5 billion survived on $2 a day or less, says the World Bank. By 2050, the world may have 3 billion more people; many will be similarly impoverished. What's baffling and frustrating about extreme poverty is that much of the world has eliminated it. In 1800, almost everyone was desperately poor. But the developed world has essentially abolished starvation, homelessness and material deprivation.

The solution to being poor is getting rich. It's economic growth. We know this. The mystery is why all societies have not adopted the obvious remedies. Just recently, the 21-member Commission on Growth and Development -- including two Nobel-prize winning economists, former prime ministers of South Korea and Peru, and a former president of Mexico -- examined the puzzle.

Since 1950, the panel found, 13 economies have grown at an average annual rate of 7 percent for at least 25 years. These were: Botswana, Brazil, China, Hong Kong, Indonesia, Japan, South Korea, Malaysia, Malta, Oman, Singapore, Taiwan and Thailand. Some gains are astonishing. From 1960 to 2005, per capita income in South Korea rose from $1,100 to $13,200. Other societies started from such low levels that even rapid economic growth, combined with larger populations, left sizable poverty. In 2005, Indonesia's per capita income averaged just $900, up from $200 in 1966.

Still, all these economies had advanced substantially. The panel identified five common elements of success:

· Openness to global trade and, usually, an eagerness to attract foreign investment.

· Political stability and "capable" governments "committed" to economic growth, though not necessarily democracy (China, South Korea and Indonesia all grew with authoritarian regimes).

· High rates of saving and investment, usually at least 25 percent of national income.

· Economic stability, keeping government budgets and inflation under control and avoiding a broad collapse in production.

· A willingness to "let markets allocate resources," meaning that governments didn't try to run industry.

Of course, qualifications abound. Some countries succeeded with high inflation rates of 15 to 30 percent. Led by Japan, Asian countries pursued export-led growth with undervalued exchange rates that favored some industries over others. Good government is relative; some fast-growing societies tolerated much corruption. Still, the broad lessons are clear.

One is: Globalization works. Countries don't get rich by staying isolated. Those that embrace trade and foreign investment acquire know-how and technologies, can buy advanced products abroad, and are forced to improve their competitiveness. The transmission of new ideas and products is faster than ever. After its invention, the telegram took 90 years to spread to four-fifths of developing countries; for the cellphone, the comparable diffusion was 16 years.

Second is: Outside benevolence can't rescue countries from poverty. There is a role for foreign aid, technical assistance and charity in relieving the human suffering of global poverty. But it is a small role. It can improve health, alleviate suffering from natural disasters or wars, and provide some types of skills. But, it cannot single-handedly stimulate the policies and habits that foster self-sustaining growth. Japan and China (to cite easy examples) have grown rapidly not because they received foreign aid, but because they pursued pro-growth policies and embraced pro-growth values.

The hard question (which the panel above avoided) is why all societies haven't adopted them. One reason is politics; some regimes are more interested in preserving their power and privileges than in promoting growth. But, the larger answer is culture, as Lawrence Harrison of Tufts University argues. Traditional values, social systems or religious views are often hostile to risk-taking, wealth accumulation and/or economic growth. In his latest book, "The Central Liberal Truth," Harrison contends that politics can alter culture, but it isn't easy.

Globalization has moral as well as economic and political dimensions. The United States and other wealthy countries are experiencing an anti-globalization backlash. Americans and others are entitled to defend themselves from economic harm, but many of the allegations against globalization are wildly exaggerated. Today, for example, the biggest drag on the U.S. economy -- the housing crisis -- is mainly a domestic problem. By making globalization an all-purpose scapegoat for economic complaints, many "progressives" are actually undermining the most powerful force for eradicating global poverty.

Mr. Samuelson's work speaks for itself. He clearly sees the problem, and identifies the ingredients of success, all of which ultimately boil down to promoting marginal productivity. He also directly asks the question: "Why aren't these lessons applied everywhere?" The answer he neglects to see is that many people holding political and economic power simply don't want to solve these problems because of their own non-productive D-brain goals (self-interests).

To be fair, he pays lip service to that horrible truth but blames the problem on culture. Culture is the result of the brain situation described in this book, not the other way round. He admits culture is hard to change. In reality it is as hard to change as imprinted brain behavior, because culture is only a mirror reflecting the brain programming activated by a particular environment. In other words, culture is merely the reflection of what the D-brain considers acceptable given its particular environment. The brain is not clay being molded by the effects of culture – the opposite is true! Cultural behavior and "norms" are created by a population's C/D-brain interactions in response to their environment. The environment can change as a result. Therefore, cultures can evolve, as brains and environments do, not the other way around.

In poor BHU-heavy countries, Mr. Samuelson is concerned about the great number of people who are being deprived the opportunity to become CHUs. But, he doesn't know who the culprits are! They are the leaders/tyrants in power who understand on a D-brain level that they must rule by murder, intimidation, and fear in order to prevent education and brain sophistication levels from rising. A tyrant's D-brain knows that if it doesn't prevent the population's intellectual advancement, he will eventually face rebellion from his own people.

In CHU-heavy countries, *industrialized nations* as we refer to them, productivity faces its own unique set of hurdles. Unions are an illustrative example of one major problem in CHU-heavy countries. For a number of D-brain reasons, unions are far more interested in union created positions, power, wealth, and confrontations, than they are in securing more productive jobs and total economic prosperity. The C-brain of the union official *believes* his behavior is in the interests of the membership. It clearly is not. This thinking leads unions into non-productive trade positions, strikes, and work place rules that strangle productivity. Once productivity is compromised, profits fall, and eventually, the union has strangled the job.

This unintentional job destruction occurs because unions want to control more than wages, they want to interfere with *how* goods are produced. They use dues to help elect politicians that further encumber the economy with unproductive ideas like job banks, import restrictions, tariffs, duties, and government regulations and safety rules that go far beyond their ostensible purpose. For example, you can't pump your own gas in New Jersey—not a safe activity, according to the unions. Unions weigh heavily in immigration "reform" designed to block lower wage labor from entering the U.S.

Unions use collective bargaining to secure things like job banks, where unemployed union members get paid to produce nothing - thousands and thousands of them at GM alone for years. Collective bargaining for wages is not harmful, but allowing unions to control how the production is carried out and how many workers are needed for specific tasks should be forbidden. As they currently behave, unions are employee wealth-distribution machines that function to strangle the production of the goods that are sold to generate wealth. Metaphorically, unions are eating the goose that lays the golden eggs. They operate under the mistaken belief that the same amount of goods and services will be produced no matter how employees conduct themselves; and therefore, any tactics to grab as much of the revenue as possible from "greedy corporate owners" are acceptable.

Let's assume for the sake of argument that corporate owners are greedy. However, since the owners are attempting to increase profits **through productivity**, they will increase everyone's wealth - if they succeed. Conversely, unions, with their current goals and agendas, inadvertently drive everyone's wealth down by decreasing productivity. *If* unions ever used their considerable *potential* power to help government and management create a business environment with strong anti-trust statutes and strong civil and human rights protections, all workers and the economy would benefit enormously.

A role like this for Chinese labor unions (predicted, but yet to come) would be a world-changing force. The current situation in the U.S., which allows unions to impact productivity negatively, simply makes entire industries poorer (witness Detroit's auto industry) and every human in the U.S. gas cloud financially weaker. If trade unions pressed for strong anti-trust laws, it would eliminate the economic disruptions associated with super-sized companies growing inefficient and imploding from their own non-productive weight. A monopolistic company's collapse crushes its employees, employees' families, employees' neighborhoods, customers, stockholders, pensioners, and ultimately the entire public.

The proposed positive efforts by unions would be instrumental in lifting worldwide wage scales, particularly at the lower end, in a country-by-country manner as this new philosophy sweeps the globe. A tide of highly efficient production protected from monopolistic inefficiencies would sweep through one low-wage beachhead after another, one country after another.

As transportation costs rise relative to goods' manufacturing cost, more and more specialty manufacturers and industries would stay behind after the low-wage manufacturing tidal wave recedes. This serves to distribute niche industries and subsidiaries more evenly, and to where they will be the most productive. Going back to the big picture: ignorant politicians on both sides of the world think they can actually control the economic gas cloud. But other than all-out war, or adoption of the Market-Fences principles outlined in this book, political alteration of an economic gas cloud is a difficult feat because of the combined laws of thermodynamics (CLOT).

Making Money is Not The Same as Being Productive...

Poor use of monetary policy can stall productivity. Remember, making money is not the same as increasing marginal productivity! Japan is the best modern-day example. During the '90's and up to 2005, *the Japanese spent 1.5 trillion dollars attempting to jumpstart GDP domestically.* Instead of the result they expected, they literally watched all that wealth disappear while the GDP growth rate barely moved. If the Japanese had simply applied the CLOT formula explained in the beginning of this chapter—which is conspicuously absent in modern economic theory--they would have gotten the result they wanted without spending any Yen.

How should Japan have used the formula?

All they needed to do was move the central bank rate from near 0% (where it had been during this entire time period). If, for example, the government in Japan announced that they were raising the bank rate to nine tenths percent on January 1st (any January 1st will do), and planned to move it again 9 months later to 1.8%, followed by one last move to 3.5% nine months after that…in 18 short months the GDP rate would be growing significantly. Why would this work at this time in Japan? The low cost of money in Japan during the '90's, robbed the D-brain driven decision-makers of their productivity motive. Why take the risk and effort needed to increase productivity when you can take money at 0% interest and invest it in safe bonds (e.g. U.S. government and corporate bonds) that yield 3% or more? Since *only* marginal increases in productivity can create wealth (increased GDP), Japan didn't realize any GDP increases.

This is a classic example of making money but not increasing wealth. The Japanese are blind to this because after WWII they funded huge increases in GDP using the cheap money formula. This worked because of a rare set of circumstances present at that time in their system. The D-brain and C-brains of the WWII survivors were *uniquely* united in purpose to prioritize Japanese job growth, factory construction, and infrastructure construction on the home island *above all other concerns.*

If you couple cheap money with highly productive behavior inside a homogenous population, you will get huge GDP increases like the Japanese enjoyed for decades. Unfortunately, this was an anomaly as a result of the massive destruction of World War II. They falsely believe, today, that cheap money alone is a sufficiently strong force to grow the economy. In reality, the post WWII cheap money policy obscured for them what really was occurring.

What a difference 50 years makes! Today, Japanese managers using their much stronger C-brains to maximize return, use the government-supplied free capital (essentially 0% Japanese Fed rate) to build factories in China and buy U.S. T-bills. China during this period in the '90's experienced GDP increases greater than Japan's and the U.S.'s combined because the Japanese-supplied money created factories for workers, which *increased productivity* in China.

The U.S. trade deficit in a roundabout way acted as another source of capital to fund these productivity improvements in China, further. So, raising the equivalent of the Japanese Federal Reserve rate would restore the productivity imperative *inside* Japan. For this experiment to work, no money can be available through governmental back doors at a lower rate, a politically difficult item in modern Japan.

The Japanese once again would be forced to turn borrowed money into industrial and service industry growth that creates jobs and generates **marginal increases in productivity, in other words: real wealth.**

This in turn would heat the Japanese economic gas cloud – expanding their GDP (wealth base). Simply borrowing at a low rate and buying another country's debt with a higher rate of return does not heat the gas cloud – despite the short term appearance of making money.

Equating the making of money with productivity is a terrible policy error. Only increasing marginal productivity truly expands the overall amount of wealth available. Money is just a lubricant in the gas cloud, facilitating the transfer of heat from point to point. Production increases, on the other hand, are almost always permanent and ratchet up total wealth and assets.

This truth can be seen in many places around the world today. Most economic schemes that make a lot of money for a particular industry or government do not do it by increasing marginal productivity. Perversely, most governments (or their agents) attempt to make money by first *limiting* productivity and then driving the distribution of the money toward a particular target. This works temporarily to enrich the target(s) but leaves less for everybody else. This resource wealth can be used to implement social policy, but not to increase overall wealth in the long term.

I am not saying that social policy cannot be used in a way to promote marginal productivity; it is just that in general, we do not use it for this. For instance, extending leave to mothers with newborns, far beyond what the U.S. does currently, would be highly productive in the long term. So would dramatically extending or expanding the funding and mandate for the Women with Infant Children (WIC) program, as just one more example. But, forcing a factory to keep 10 men more on a production line than is needed to perform the job is just destroying those jobs *and the ones next to them.* "Make-work" policies are a corruption of the economic system, and usually decrease wealth for society as a whole. Loss in productivity, like a union job bank, has the same negative experience on total wealth as graft or corruption. It's just a legal form of the same negative influence/cooling of the gas cloud. Most societies do not understand this, even in the industrialized countries.

Poor countries have large numbers of BHUs clustered around a relative few CHUs. Industrialized nations have large numbers of CHUs and relatively few BHUs. A cloud with a lot of BHUs takes longer to heat up because the productivity multiplier atop each BHU is so much smaller. One BHU can be advanced to make twice as much...in essence making a double BHU. When productivity gains in that original BHU are increased to three or more times the original subsistence level, he/she becomes a CHU.

As stated before, this only happens slowly, as a BHU must leave his subsistence comfort zone through education and effort in order to generate three or more times what he/she and the family needs. As this point, the BHU becomes a CHU. A sufficient concentration of CHUs allows the productivity multipliers to increase exponentially through productivity-enhancing interactions. In a modern CHU-heavy economy, we can disregard the BHU contribution and abbreviate the total wealth equation to:

Average CHU Effort (E) x Aver. CHU Prod. Multiplier (P) = per capita GDP.

This IS the heart our economic gas cloud. E x P = GDP. Period. Increase effort or the productivity multiplier (through education, capital, etc.), or both, if you want the cloud to expand! Decrease either, and the economic cloud contracts. To transition many BHUs into CHUs requires capital, political stability, private property protections (this generally takes the form of land reform) and education for the majority of the population. It takes a certain minimum amount of these elements, and productive use of time, for these to be applied and add heat to the overall gas cloud.

Mr. Samuelson blames the lack of these necessary ingredients in the poor areas of the world on culture. This is inaccurate. The reasons can be found in brain evolution. There are evolutionary reasons that impact a population's ability to create a rational plan for advancement.

The economic road forward for these impoverished populations initially includes **limited** power-sharing with their rulers, a focus on education, some property rights (land reform again), and productivity connections to the rest of the world. These connections must include targeted trade tariffs and duties, and infrastructure improvements to the transportation network, particularly roads—the greatest physical productivity multiplier of all time (not counting the printing press). The targeted trade tariffs and duties (plus Market-Fences in an ideal situation) function to force the nascent CHUs to compete with each other in an environment safe from powerful external forces that would eliminate them.

Once the CHUs industry is healthy and productive, the tariffs would be reduced or eliminated (and the Market-Fences adjusted) to encourage worldwide competition. This in turn would fuel more competitive behavior, and thus more productivity advances for all the CHUs in a particular industry. To lift their people out of poverty, a BHU-heavy society/nation needs to transition their BHUs into CHUs. If this does not occur, sustained improvements in standards of living/GDP simply cannot occur.

Leaving things the way they are, would be labeled "Imperialism" by the tyrants in these BHU heavy nations. This means having a social policy that evolves the BHU emerging C-brain, through education and property rights, into a C-brain strong enough to compete with the D-brain forces arrayed against it -- both domestically and internationally. This requires, at a minimum, a C-brain that can read and write, and work productively with tools at greater than a three times subsistence level.

The above knowledge has large foreign policy implications. Installing a democracy in a country without a majority CHU population is a fraud.

Imposed power sharing, forced upon a non-power sharing inclined people (D-brain heavy BHUs) by an economic or militarily superior population is guaranteed to fail. Education and infrastructure must precede a power-sharing government, or it will be neither self-sustaining nor productive. Once improvements in education and infrastructure lead to a rising and sustainable GDP, the people themselves will decide their desired level of power-sharing with the government.

God is love

Book II, Chapter 4

Effective Anti-Trust Laws: the Economic Holy Grail

Although Anti-Trust is one of the most important principles in economics, it does not get the attention it deserves. In part, this is due to the poor manner in which the necessity of anti-trust regulation has been explained to the public. More importantly, the existing anti-trust statutes do not fully express the principle upon which they must operate, and hence they are not fully effective. To benefit the human economy going forward, this must change.

Simply put, the vertical and horizontal organization of our economic and political infrastructure must be modified, creating a new economic model. This new model will stimulate the highest possible level of competition in each individual market segment by constructing "Market-Fences" around pre-determined market segments, or "Segmented Gas Clouds." Each fenced area would represent the sum of that market segment's consumers, distributors, manufactures, and all affiliated industries necessary for the proper functioning of the primary "cloud" -- e.g. insurance, investment, banking. All the market segments taken together would fit together like pieces of a jigsaw puzzle, adding up to the total economy.

To maximize our chances of surviving and thriving as a species, we need to use the productivity multipliers available from C-brain logic while harnessing the competitive drive of the D-brain. By using both parts of our brain, we can *rationally* provide for managed competition. In physics, as outlined in the previous chapter, we call this an ideal gas. In other words, the gas molecules depending upon their type, temperature, etc., either exist in this ideal state and exhibit all their distinguishing characteristics, or they do not because one of the forces, such as temperature, pressure, or volume, is out of balance. If this occurs, the economic gas cloud stops acting as an ideal gas, and the individual gas molecules lose their distinguish-ability.

This is what limits marginal productivity in the version of the gas cloud where humans are the gas molecules. The real purpose in designing Market Fences is to identify how to create monetized economies where humans compete to produce food, fuel and all the derivative products ever more productively. This state of heightened productivity can only be achieved when *all* the characteristics of an ideal gas cloud and pre-constructed Market-Fences are present. This is also the most efficient way to do this because each market segments' objectives, societal restrictions, priorities, taxes, and grants, throughout the segment, are open knowledge. Everyone in the market segment can participate and compete on even footing with the sole advantage

The Revelation

going to the most creative and efficient producer of goods and services. Market-Fences would combine several legal and regulatory concepts not integrated today. One force heats the cloud by greed: enabled through the C-brain denial of risk. Another cools the cloud because of D-brain generated fear. Typically, market instability in competitive economies is the result of one of these forces being out of balance. This instability naturally results from the dynamic struggle in the C- and D-brain structure of humans.

How do we prevent these apparently unavoidable fluctuations from impoverishing us while they cleanse a particular segment of unproductive practices/companies/entities? Remember that economic gas clouds must follow the Combined Laws of Thermodynamics. Therefore, any aberrations will eventually self-correct despite our attempts at manipulation.

Unfortunately, the laws of thermodynamics have no respect for the harm to individuals that such corrections can cause. So, it is better to set up a system that anticipates and harnesses these economic changes in the first place. The description of the output of the economy, GDP, is simply the product of all the work done (effort of the individuals) multiplied by the productivity factor. The behavior of a healthy economy, or segment of the economy, can be expressed by a simple mathematical equation adopted from Newtonian physics:

Work (**force**) x Productivity (**d**istance) equals total Wealth (**E**nergy).

Market-Fences could eliminate entire national economic market swings up and down by confining the corrections to a small number of "fenced" market segments at any one time. In a well-developed Market-Fenced economy, a majority of segments should be moving up, as other segments are moving down. The stabilizing effect is the result of the cleansing force of competition within each segment, allowing the whole economy to proceed upward through expansion and increased wealth for all involved.

Absent Market-Fences, the underlying economic segments (gas cloud), can easily be influenced by multiple C- or D-brain driven unproductive forces. These forces serve to destroy the competitive environment within "unfenced" market segments, thus deteriorating economic health (cloud cooling - negative GDP influencing). It *should* be obvious to everyone that it is self-destructive to let companies, governments, unions, investors, special interests, and/or just rich people distort a market segment for short-term personal gain. Yet, through bribes, large campaign donations, lobbying, and various inducements given in order to make a company, product, or cause (deregulation/ethanol etc.,) triumph, this self-destructive distortion occurs, albeit briefly, through the creation of anti-competitive forces that favors them.

Inflation and the Gas Cloud

Inflation is the result of a rising expectation by individuals and companies (the gas cloud molecules) that they can get more money for the same goods or unit of work. This increase in payments to individuals can only happen without inflation, putting aside the issue of savings and debt, when the molecules/humans in the gas cloud are producing more material and/or service per unit of work than they had been previously (a marginal gain). As we can see from the above equation, this can happen only through increasing the productivity multiplier and/or total work done.

Inflation can also result from the imbalance between the amount of liquidity in the gas cloud and the supply of goods and services being produced. In this case, excess money or liquidity either from external unearned sources alone, or coupled with slowing productivity, can cause inflation. The unavoidable result during the correction of inflation (assuming productivity does not dramatically rise) is deflation and recession, as reimbursement for products/services decrease to the level appropriate for the productivity being generated. The United States' 1970's stagnation in productivity was the main cause of that period's inflation, not the "historically" stated causes favored by economists today (the Vietnam War competition-for-resources theory).

If you don't have Market Fences protecting segments of the economy from the destructive influences created by those who wish to gain an uncompetitive edge, these harmful forces will undermine the productivity factor making work output less efficient. This inevitably decreases wealth production. It should be clear to even the casual reader that it is incredibly destructive to a nation to allow companies or people to influence the means of production in a manner so as to make it less marginally productive or competitive. Fortunately, the opposite is also true.

To create an expanding economy, a rising stock market, increasing standards of living and a content population, all government needs to do is create an environment where increased marginal productivity is encouraged, protected, funded, planned for, and mandated by governmental laws and regulations through the use of Market-Fences. It is truly that simple.

Here are a few examples of Market-Fence applications from varied points of view: a successful "Fenced Market," an environmental disaster in the making, and an industrial behemoth, crippled by a poorly designed fence.

The NFL. Supposedly an "Anti-trust" violation itself, the league is actually well structured to provide for controlled *athletic* competition. They control the number of teams, keeping the number high enough to have national appeal, but not unlimited so the league loses uniqueness. They balance talent, coaching, new stadiums, TV rights, draft picks, and rules every season -- in order to keep the competition at a high level. The result is that American football players have literally evolved right in front of our eyes. A team from 20 years ago, if transported in time, would perform very poorly against a modern NFL team because of inferior weight, speed, strength, quickness, and vertical leap. Today's players are bigger, stronger and faster. Productivity has skyrocketed!

The Oceans are a giant, largely unregulated market segment necessary for globalization in the form of wealth enhancing trade, and simultaneously serves as the supporting base of the world's entire food chain and ecosystem.

Despite the importance of the oceans, humans have not united for the purpose of preserving this critical resource. In fact, no one nation, no one entity, no one political or military force, controls the oceans. As a result, at the current accelerated rate of species extinction and stock depletion, we will collapse the ocean's biosphere within the next 30 years. That could be our demise as a species.

This is a species-survival crisis both for ocean life and land life. The environment can be expressed as a mathematical formula that begins with the energy of all the sunlight hitting the Earth every day. Sunlight is the original source of all energy that powers Earth. The end results of the chemical reactions powered by sunlight are the products that we trade in our monetized (and non-monetized) economic gas cloud. Even our individual human effort, except for the quantum energy-God part we will discuss in the last part of this book, is wholly derived from the sun's energy stream. The most significant portion of this energy stream, the roughly 70% that hits the oceans after penetrating the atmosphere, provides energy to Earth's life cycle in the top 100 to 200 feet of the ocean. To interrupt the oceans life cycle creation through species depletion and pollution in this critical ocean layer is to threaten our own survival. We are doing exactly that. This is not an exaggeration of the ocean's (and therefore the Earth's) environmental problem. My suggestions: The oceans are in **dire** need of coordinated protection, if only mankind could reach some agreement here at least. In the meantime: Oppose dumping waste into the sea, *including* farm, storm-water system, and temperature elevated run-off. Oppose all commercial fishing until the oceans recover and can support wild-species-safe, managed fisheries/aqua farms. We have little choice. We're assisting in our own rapid demise if we keep killing the ocean's life.

Automobiles. Look how healthy, profitable, diverse, flexible, intelligent, competitive, and highly productive the Japanese auto industry is compared to that same market segment in the U.S. The cause is simple. It's not about subsidies. It's purely about competition. Two examples:

One - Japanese workers do not fight productivity improvements that mean fewer workers per car.

Two – Japanese auto companies grew up in an economy protected by a type of Market-Fence inadvertently designed by the Japanese government with input from the U.S. economists that helped Japan rebuild after the war, including Harold H. Thomas, a market genius. The U.S. hasn't had real internal competition in the American auto industry for 40 years, and the health of our car companies shows it.

U.S. economic trade, environmental, and anti-trust policies toward the auto industry for the last 50 years were inadvertently designed to shield the American companies and their workers from the need to improve marginal productivity. In essence, what the corporate, government, and union leaders did was "cash-in" the productivity lead the American auto industry had at the end of WWII for falsely elevated values in worker salaries, stock dividends, and management pay.

In further proof that you cannot change the laws of physics as they pertain to the economic gas clouds, the unavoidable result was to make U.S. auto manufactures weaker. The end result is that they are bloated, debt-ridden, inefficient, and tied down by protectionist-seeking management and union contracts that make managing the companies nearly impossible.

It gets worse. Ultimately, the auto industries' pension obligations will become a national problem when these failing dinosaurs breathe their last. They could become great companies, particularly if they were broken up into *smaller* companies and operated under rules that encouraged/allowed productivity gains. Improve productivity every day, in every program, from every worker, in every design, and through every deal. This is the only way to salvage these companies. If they improve their productivity, even these near dead companies and most of their workers will prosper again. The auto industry is no different than any other industry of elaborately manufactured items. All market segments will thrive, if competition is encouraged, incubated, and never allowed to consolidate.

> In essence, each fenced market segment needs to be protected from the forces of the human D-brain. The D-brain does not want to compete, exhausted to the last man on the field. Such competition, though enriching to mankind, is too risky for the D-brain. The D-brain says: "Give me security; give me a monopoly, give me union benefits and tenure; give me welfare and job banks; give me unconditional security." Only this wealth-destroying guarantee of unconditional security calms the D-brain's obsessive fear of death, need and want. In a cruel irony, it is this D-brain obsession with security that creates the conditions that ensure our boom-to-bust economic cycles.

Michael Crichton described this as a mass failure on the part of humans to perceive the difference between different levels of risk, causing us to focus on things that pose small statistical risk while ignoring the obvious larger risks. Dr. Crichton sees the importance of this from an efficiency point of view. In this regard, he is correct. If we spend millions saving humans from things that aren't likely to harm us in great numbers, by definition this uses resources unproductively. These resources would otherwise be capable of saving hundreds or even thousands of other humans from significant risks that we don't perceive as a threat, even though in many cases they are more dangerous by far--like automobiles, which kill 40 to 50 thousand people each year in the U.S. alone. Half of those deaths could be completely eliminated with drunk driving lockout devices.

What Dr. Crichton failed to perceive is the D-brain reasons for these inequities, which override the C-brain logic Dr. Crichton so eloquently displays. This obsession with small risks to the ignorance of large risks exists in many areas of the economy other than simply medicine, health care, and automobiles. The D-brain, which assisted in survival for millions of years by fearing the unknown, still fears the unknown.

This occurs even in a universe in which man's C-brain has practically eliminated the primitive causes of the unknown. This inevitably leads us back to the core of the argument for humans to develop economic Market-Fences: to control the corrosive power of greed and fear through new economic institutions and laws. Only in this manner will we maximize productivity through competition.

Increased Productivity Through the Use of Market Fencing

Increased productivity equals increased real wealth. Decreased productivity and wealth occur when:

A. The cloud does not have enough liquidity to transfer the heat efficiently. (1930's)
B. The cloud has too few members (a monopoly) so it behaves non-competitively.
C. Patent law abuses occur (e.g. pharmaceuticals).
D. Tax treatment inequities block competition.
E. Work place rules exist that don't marginally increase safety or productivity while decreasing productivity through inefficiencies.
F. Any other conditions, real or legal, that disturb competition and efficiency will change the future negatively for all of us.

Energy Crisis and the Death of Humans

Twenty to 30 years ago, in the market segment of solar power, a Market-Fence was needed to preclude corporate ownership of patents, companies, and production facilities and/or stock by any company possessing an ownership stake in a fossil fuel energy company. This reasonable type of restriction would not have been a competition-killing regulation. The net effect would have been the opposite; it would have prevented the oil companies from interfering in *other industries* with the goal of preventing competition to their oil dominated energy markets someday in the future.

We need incubating Market-Fences around industries developing solar power, wind power, biomass and geothermal power, and for some as yet to be discovered technologies. Legislative Market-Fencing would have prevented the energy crisis the world faces today, a crisis that will kill millions before it is finished. How? The "oil shock" of the 70's encouraged some innovative people to think about where oil scarcity might end. By the early 80's a small cottage industry had developed of independent companies pursuing the production of photovoltaic cells (material that converts sunlight into electricity) at a cost effective return ratio. Currently it's 50 to 100 times as expensive to make a photovoltaic energy cell, as it needs to be. Why? Because the big oil companies purchased all the start-up photovoltaic companies within a very short period of time and essentially shut them down with only a few pathetic exceptions. This is shocking but correct! The U.S.

government *allowed* big oil companies to buy up the entire photovoltaic industry—almost overnight--less than 20 years ago.

Most of these alternative energy companies were then shut down. This is D-brain corporate thinking by fossil fuel companies, simply killing the competition. This denial of the human race's access to an inexpensive energy source will kill many millions of people and could kill billions of people. *When honestly viewed for what it is...this is reckless endangerment at the least or premeditated murder at the worst!* I say this because the photovoltaic industry was on the verge of two huge breakthroughs, both of which have been proven scientifically, neither of which has been put into production.

First, high efficiency (15%) photovoltaic solar cells can be *extruded at a production-cost to energy-generated ratio that threatens the use of oil.* Bell Labs extruded silicon wafers *more than 40 years ago.* I am not referring to the Czochralski growing process method employing diamond coated wire saws which is also termed extrusion, but to the flat extrusion of a silicone wafer suitable for doping *without* being sawed. Bell Labs removed the ceiling tiles and struts in the lab to complete the experiment. I have spoken with a scientist who was there as they "pulled" and extruded a silicon wafer up out of a vat of pure silicon. They were looking for different ways to make computer chips. Instead, they actually stumbled upon the solution for photovoltaic production but didn't realize its value.

BP Solar (formerly Solarex) is also capable of extruding silicon wafers. Even though they aren't marketing solar cells made this way, they fully understand how to manufacture them using this process. They simply don't put this technology into production. In my opinion, this technology has been known to every major company in the oil industry for years but has been suppressed because there is too much money at stake to allow the solar cell industry to blossom. If society had allowed the people who sold ice out of the back of a wagon to own the refrigeration patents, we would all still be hauling ice into our businesses and homes today. The big issue here is, what they did was not illegal. Our laws allow this type of activity. So, it's not surprising from a D-brain perspective, in fact it is predictable, that fossil fuel companies would stifle alternative energy source production.

Defining Market Fences

The definitions of what is included in each Market-Fence should under go expert review on a regular basis. This serves two purposes:

1. To fine-tune the definitions and concepts behind those definitions so

that they keep pace with the technologic and market conditions present in that market segment, and

2. To determine if any tax breaks, low interest-loans, and/or tax levies and surcharges are needed to either stimulate or constrict activity in that segment.

Trade tariffs, tax incentives, low interest loans to stimulate activity, grants, market definitions, societal goals and knowledge from past lessons--a Market Fence is where all these elements can come together. This would give the investor, the business owner, the consumer, the citizens of the world and those entrusted with protecting our environment a way to discern what combination of elements create the greatest productivity increases (hence, the greatest wealth available to be distributed) without violating the long term health of the industry (market segment), the environment, or the consumer.

This system would also cut down or eliminate productivity-choking over-regulation from overlapping authorities while simultaneously only allowing competitors that have the potential (motivation) to optimize the segment within the fence. This is the way most effectively to promote a gradual heating of the gas cloud with the least loss of human life, the least disruption to the business cycle, and the most balanced and effective distribution of wealth to the greatest number of people.

The reasoning behind this is straightforward. If we want to increase productivity to the levels necessary to prevent starvation, famine, epidemics, social and economic unrest, war, and even significant depopulation events we simply need to apply these lessons in law, social custom, and/or in regulation. These are critical places where the interplay between the human mind, the nature of the economy, and the physical laws of the universe affect productivity.

A pessimist might say that since the human brain is good at lying and denial, you cannot count on man's adherence to the rule of law or his good nature to protect the consumer. This is true. An optimist would say a "good" man is concerned with feeding his family–so he naturally will protect and/or seek a monopoly or exploitive advantage if he can get it. This is also true. This means that only through controlled competition (that cannot be manipulated) will healthy profit levels be generated, promoting additional productivity increases while not simultaneously over burdening the consumer.

Since business owners will leverage every existing variable to optimize their business, market share percentage must be one of the key variables limited by the Market-Fence. This would vary widely from market segment to market segment. Obviously, in the aircraft-manufacturing segment you will not have fifty competitors, but in TVs you might. Market-share percentage is an absolute. If one company's market share gets too large, all will know this. Keep the size of any one company beneath 40% market share, the number of companies in each market segment numerous, and let competition restrain most excesses not prevented by other elements of the Market-Fence.

What is Competition?

We must begin by coming to an agreement on the definition of competition. Under the economic and political philosophy of today, free markets are those, which are allowed to operate with a minimum amount of regulation and interference from the government. Communist and/or socialist economies are those that operate with the most interference. As this book explains, these labels are not accurate. Communist societies are usually non-power-sharing versions of tyranny. Because the early power-sharing democracies adopted free enterprise style capitalism, the people in those societies integrated those values. In other words, democracies integrated free markets into their *political* and economic framework. China today, completely exposes the flaws of our old definitions. A communist leadership, with complete political authority, has allowed capitalism to be their economic engine of productivity with the resultant GDP increases that are nothing short of miraculous.

The U.S. has done its part to expose the flaws in those definitions as well. People supposedly do not like government interference in the markets by politicians and their agents because market forces are supposed to be strong enough to self-correct. Competition is a D-brain generated motivator. It's part of your human defense and production system. This is not a virtue, despite what Milton Freidman and Adam Smith-style capitalists think. Since competition for "deer meat" is a powerful base motivator, it naturally will run amuck in a C-brain world in which restrictions on its D-brain desires are abundant but ineffectively designed.

Take, for example, 1929 and 2008. The politicians, desperate to save their people's faith in democracy, used the powers of government to rescue the banks before everyone withdrew their money, which would have caused the system to collapse completely. People believe in the free market, in competition driven productivity, but only believe it in their C-brains. Threaten their pile of "deer meat" and they demand regulation for protection.

Some C-brain schemes, like trying to make 80,000-lb aluminum airplanes fly, will work if they are based on underlying scientific principles. Fortunately for mankind, anti-trust is one of those types of schemes. Anti-trust is a C-brain scheme, which runs contrary to most D-brain thinking. This conflict results in most anti-trust laws being poorly written and poorly enforced. The C-brain "knows" it is needed, but the D-brain doesn't like it. By changing anti-trust laws through the creation of Market Fences, we can harness the power of competitive companies, without the boom/bust cycles caused through cyclical greed and fear.

Each time there is an economic catastrophe; D-brain fear drives the decisions of the masses. Ultimately, this causes financial panic when something in the system -- e.g. margin lending (1929), derivatives trading (2008), tax law changes (1986) -- goes wrong and the economy begins to shrink. Once this happens, only a government reaction can stabilize the situation. The government has the powers of life and death over its citizens. This is the only thing the D-brain fears more than losing its money. The day the Brooklyn Bridge opened, somebody screamed. Somebody else thought the bridge was falling and began to run. Many died in the resultant panic. The bridge never moved an inch; it was built much too strong for that to happen. Crazy with fear, humans were running and trampling other people while they tried to escape from an irrational fear. Humans, who moments before were strolling in their Sunday best while a band played sweetly in the background, became beasts on the stampede to escape their fear.

Contrary to public opinion, the majority of economic crashes do not occur because Wall Street steals too much, though I am not defending them. The majority of the time, economic systems crash when millions of investors, spurred on by herd mentality, build a bubble that is destined to burst. Regulation, as it is currently understood and used, fails to prevent these bubbles **and** simultaneously damages the total amount of goods and services produced by creating biased competition. We therefore need first: a new definition of competition. Then a new definition of risk: and finally, new types of regulations that incorporate these definitions and our new understanding of anti-trust principles to prevent these harmful economic events from occurring. This new, much more extensive, type of anti-trust tool, which combines some of the legal principles in existence today *plus new ones*, should be created to form what Dr. Richards and I have labeled Market-Fences.

Market-Fences

Many U.S. federal judges will not enforce the anti-trust statutes as they stand. While understandable, this is a great economic tragedy. A strong set of anti-trust laws based on the principles outlined in this chapter would enormously benefit the U.S. and the rest the world with an unprecedented burst in productivity. We *need a burst* of economic expansion to get past the age of fossil fuels, and to allow us the opportunity to produce and distribute sufficient food/fuel to an increasing world population. I was going to write a separate chapter on oil, one on solar power, and this one on Market-Fences; however, since the examples and principles are so intertwined it seems best to combine them into one. The ideology as presented here, is a lot bigger than traditional anti-trust regulations. But, the Revelation says it will work and is necessary for our survival.

Economies, as we previously discussed, *behave as gas clouds do*. We want the molecules in that cloud to be able to move around freely, bumping into each other in order to transmit heat, but not so compressed as to lose their distinguish-ability. When the gaseous material of the economic cloud itself, humans and assets, is too connected to one company, one idea, one philosophy, one entity, one method of production, one level of productivity, or one pay source, then productivity improvements and output diminish or disappear from a lack of competition - and total wealth begins to decline. There is less bumping occurring, and less heat being generated or transmitted.

Once the gas cloud loses that critical component of distinguish-ability, the humans in it can no longer distinguish between the more or less-productive activity. Since there can be no selection of better choices, productivity stops increasing and eventually starts decreasing. Once distinguish-ability disappears, investor confidence also falters, credit markets collapse or shrink, and productivity decreases become systemic. We used to call these crashes; now they are either classified as recessions or depressions in the industrialized world, depending upon severity.

Humans must be able to distinguish between the more or less-productive elements of the economy and flag this data or **highlight** it. Otherwise, D-brain programming (net sum logic) will select options not necessarily based on productive, logical, or sustainable choices, driving our combined behavior toward less productive solutions that cool the economic cloud in which we live. This occurs on hundreds of levels in the combined gas cloud of the world economy, even in a regional economy. The more distinguish-ability is restricted, reduced, or eliminated, the more vulnerable the system is to crash. On this point, Paul Krugman, a *New York Times* columnist is right on. He argues that it was the process of allowing mortgage debt to be sliced up and repackaged, and then sold, what the experts call Collateralized Debt Obligations or CDOs that is the one of the unsolved problems at the heart of the 2008 financial crash. He is absolutely correct. The change in the system to securitization that he describes in his columns (he correctly argues we must abandon securitization) is a disturbance to the proper role of risk in the gas cloud. Nothing the Obama administration does concerning these loans will work to long-term success if the policies don't begin by attempting to utilize risk as a tool rather than just an unpleasantry to be eliminated. This means the end of CDOs and derivatives, by *force majeure* if necessary, and it should.

If, after analysis, it is found that humans are not adequately deterred from speculative buying activity, property abandonment and or over consumption in housing square feet unless a 10-20% cash deposit is required for every residential loan, then set the limit there by a process recognizing the need for this balance rather than a political process tied to the government's wholly unrelated concern of financing the nation's debt or boosting homeownership past sustainable percentages. Risk is what ties the homeowner's productive labor to the property.

Society should embrace transparent risk as a lever to assure the safety of society's combined excess productivity (money in the form of a construction loan) used by developers to create houses. Any other strategy puts society's combined excess productive capability at risk! In other words, if the government's need to push public debt (to finance their activities) is tied to the financing of home loans, then home loans won't be made on the basis of restricting speculative human behavior, but rather *fueling* that behavior. In this regard, such a common-sense rule (10-20% down), which actually existed for decades, would have saved the government from the damage caused by the government's own lack of knowledge about what was happening in the gas cloud. No income limits, no asset type mortgage loans will inevitably result in this problem every single time. While this was occurring at the bottom of the credit market, derivatives were weakening it from the top.

Think of the 2008 market crash when the total derivatives market exceeded world GPD by a factor of 10 to 20 times. These products, ostensibly designed to spread risk, actually lumped all bets, good and bad, into indistinguishable baskets. Risk is good. Try to think this way. By its nature, it selects between uses of capital that are productive and those that aren't. Spread that risk and we lose the indicator between good and bad uses of capital, between good and bad assets, between good and bad risks, good and bad choices, good and bad projects, virtually everything at the end of the day.

Obscuring the function that risk performs, a goal of government and financial experts in the 1990's, is actually a big mistake. Financial markets should never allow the investor and lender too much distance between them. Once this happens, the entire cloud must cool. It cannot do anything else, mathematically or in the real world of Wall Street; because without finely tuned productivity indicators showing us the way, our D-brains dominate our decision-making and, driven by net-sum programming logic, instinctively choose marginally less-productive outcomes.

Furthermore, the interconnectedness of large, monopolistic, "un-fenced" entities spreads decreases in marginal productivity as if it were a disease or virus. A visual example: Imagine a series of watering tanks fed by a stream that runs through many different pastures. Poison the stream, and it affects every creature living in every pasture. The massive amount of derivatives trading on world financial markets actually did exactly this in the 2008 meltdown. This one stream, the largest ever created, poisoned many water tanks in many pastures, even though the argument behind creating this investment vehicle in the first place *was the desire to lower risk by spreading it*. The goal of lowering financial risk by spreading it is specious at its center. In fact, it's almost always a bad idea. Things that fail, distinguish for all observers what not to repeat. This process is necessary to continue improving marginal productivity. Overall human wealth cannot be increased without doing that. Spreading the risk *obscures failures* and makes it difficult to recognize success. Unrecognized failure(s) are allowed to continue, lowering the wealth level of every person on Earth.

The reason/force behind this phenomenon is the power of the D-brain. If you give a corporate executive enough money and market power, he will abuse that situation by attempting to use money and power, through lobbying and manipulation of the market to maintain an elevated position. (While unions are correct as regards corporate management's greed, their current solutions worsen the situation.) More importantly, destructive human D-brain behavior is magnified by the herd behavior of consumers. Since humans are herd animals, we tend to reward the biggest, the best, the winner, the hero, "our" alpha. Our brain tends to reward the "biggest" with customer (brand) loyalty. This consumer "biggest brand" phenomenon works against the creation and maintenance of optimum sizing for corporations.

CLOT generates the highest levels of sustainable wealth and production only in market segments where the gas cloud is composed of elements that are in balance with each other—an ideal gas that is not too dense and not too empty. Hence, a major goal of each Market-Fence overseer/regulator should be to keep the elements (in each segment) in the right relationship to each other. These would include relative firm size, access to capital and access to educated, motivated, capable employees. It would also extend to physical resources, environmental, and other regulations, including taxes and/or tariffs.

Quantum Mechanics in Economic Thought

Think of the economy in the following manner. Pretend that each time you make a transaction with money, by trade, or by agreement, a gas molecule in a cloud bumps into another gas molecule. These two molecules trade information and energy, *and* trade a metaphysical "particle" at this moment. These transfers are not just two-way. The information transferred has a quantum (particle) element that connects it with every other molecule that touches the molecule that you have just touched. Put another way, every transaction you make simultaneously affects every other molecule in the economic gas cloud. Please re-read this sentence because this concept is not intuitive or obvious. (For more on this concept, please read the brand-new book by Bruce Rosenblum and Fred Kuttner, entitled *Quantum Enigma*.)

For example, when you buy a tank of gasoline at an Exxon station, that transaction, in the assumed amount of $100, has $100 of initial gas molecule collision value. By definition, this must be the same for each colliding molecule. Double entry bookkeeping is the mathematical translation of this core truth. Whether the good exchanged is worth the face value of the transaction or not, the "label" on the transaction is initially equal. The transferred good, on one end, e.g. 28 gallons of gasoline -- from one storage inventory to the other – is equal to the corresponding money transfer on the other end.

Money, then, is just a symbol of value. It represents both the value of the goods and an amount of work the purchasing human did. Accountants long ago discovered the need for "good will" as an accounting item, to make sense of this metaphysical property. Unfortunately, this device, since its origins were not scientifically understood before now, has been abused on corporate balance sheets for decades.

It is our human belief in money, and in the monetary value of work humans perform, **plus** *the concentrated energy transmitted from the Sun, that gives the world's economic (gas cloud) system the energy it needs to run.*

Potential Wealth: An Example

A forest has a lot of valuable lumber. Un-harvested, it falls to the ground and decomposes, releasing its carbon into the atmosphere. Harvested, it is energy/potential wealth that can fuel heat for the cloud. It is potential energy, translatable into potential wealth, given to us from sunlight through photosynthesis. The harvested lumber is used by BHUs for heat, or used by CHUs as a salable commodity, which is where the wood joins the monetized portion of the economy as inventory.

The Physics of Good and Bad Faith Transactions and Uncertainty

During a transfer of goods and money, the quantum part of the transfer can be degraded or subtracted from the total transfer if there is a loss of reputation or confidence associated with the transaction, (negative good will). A CHU, represented by a gas molecule in our economic cloud, could make 100 coins worth of product but could conceivably do 200 coins worth of negative wealth damage to his reputation from a transaction. Before long, after enough of these negative transactions, he will not be able to interact with other molecules since they view him as a poor value and will avoid him.

In this way, the total energy of an economy reflects the uncertain nature of the human relationships. Although the total energy of any one molecule can never be exactly known, you can estimate the potential contribution (energy) level of a BHU or CHU (molecule) by the behavior of the molecules surrounding it. (For those familiar with the Heisenberg Uncertainty Principle, this should be a Eureka moment). Because of uncertainty and the potential for negative interactions, the more interrelated the molecules (BHUs, CHUs, and Businesses) in the economic cloud, the greater the danger of systemic failure. This potential for negative results must be taken into account much more than we currently do.

We can make our economy much more successful with significantly less *overall* risk by simply applying the principles of anti-trust and Market-Fences. Optimally, we need to act and create wealth locally based upon information shared globally. Globalization on a scale that truly benefits mankind cannot rely on a few vast factories making the entire baby formula supply, or whatever product or service, for the world. This just looks efficient to the D-brain. Rather, mankind is much better served by having highly competitive smaller factories dotting the globe, all using the latest techniques, information, and knowledge about milk and formula production, *information* shared globally.

Production failure and/or safety concerns at even a few local factories would still allow sufficient formula to be supplied worldwide. Deaths or illnesses would be contained, and best practices' accountability at every factory worldwide would be intrinsically better because competitive companies generally produce better products and services. And; their products or services would be continually refined, changed, improved and ultimately made safer and more efficient. We have raised worldwide GDP/per capita by 65 times in 500 years using much of this formula. This leads to a second reason for the need of Market-Fencing. Fencing, for example, would prohibit ownership of a small solar energy company by a large oil company. In addition to the ideological conflict posed by an obsolete technology corporation owning the intellectual rights to its emerging successor, there is the very real problem of a corporate giant acquiring a valuable company that is so small that its intentional demise is not a significant economic loss to the acquiring entity.

How our Oil Addiction May Kill Us

Our D-brains used to *regularly* figure out how to feed the family when the hunt failed. This notably included attempts to kill others who were successful, in order to *steal* their deer meat. In modern times, this behavior was exemplified by the Teapot Dome scandal, Standard Oil of Ohio, Enron, World Com, etc. As the enormous leverage of globalization and advancing technology increases, the end result of continuing this type of behavior will be the death of *billions* by starvation. I said Billions. I am not a doomsayer, nor am I a conspiracy theorist. However, I believe conspiracy "type" results are possible by the implementation of our D-brain impulses using our C-brain manipulation knowledge, without the use of secret meetings or tacit agreements. The Revelation engendered little respect for the way our species handle our natural and artificially created resources.

Oil is currently the main source of the energy we use to monetize our work and create GDP. While human work is the first productivity multiplier applied to utilize the energy from the Sun, past and present, oil is the muscle that facilitates *most* of the remaining work. During the 1950s a scientist working for Shell Oil, Mr. Marion King Hubbert, predicted the world would run out of oil one day. This prediction may seem obvious to you and me and anyone who has ever emptied liquid out of a container that is finite in size.

Remarkably, the world has done little to prepare for that day. Mr. Hubbert plotted a bell- shaped curve to predict the very year we would begin to run out. He charted our oil supply in a graph showing the increase, the decrease, and the eventual depletion date. This graph is so fascinating; I've included a copy on the next page. It predicted 20 years in advance, that U.S. crude oil production would peak in 1970 (it did in 1971). It was a prediction he was professionally ridiculed for, even though he was later proved right.

The second graph is worldwide oil production by country. Notice how the two graphs resemble each other. This Department of Energy graph was temporarily available on the U.S. Department of Energy website, but has been subsequently taken down. As you can see from these two graphs, Mr. Hubbert knew exactly what he was doing. This is true because Mr. Hubbert modeled the graphs he plotted on production numbers he saw coming from active oil fields. He then developed a mathematical model, which he used to **predict with accuracy** what would eventually represent the entire industry. As mentioned, his model predicted, *almost* to the year, when oil production in the U.S. would peak and begin to decline. His *worldwide* oil supply numbers were slightly less accurate owing to a number of external events he didn't predict, such as OPEC, the oil shock of the 70's, and the discovery of two or three new major deposits. Nevertheless, he was right enough. Over the course of the next ten to twenty years, humans will endure the dramatic effects caused by the end of the Age Of Big Oil. The first graph is Dr. Hubbert's prediction; the second graph is our actual usage and reserve history according to the U.S. government.

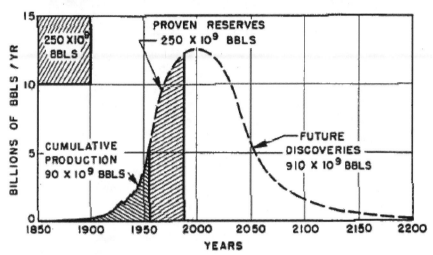

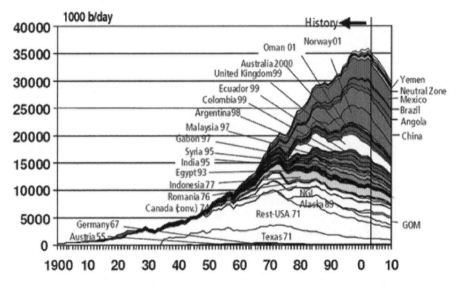

So, if something -- for example oil companies -- interfered with the development of innovations and/or alternative sources needed to replace that energy supply, they would be tinkering with the actual power source that keeps you and billions of us employed, fed, and warm. That is exactly what has been allowed to happen. Mature technologies such as nuclear power and coal can buy us time, but have significant negatives. Ultimately, these negatives translate into efficiency deficits for our economic cloud.

Therefore, more productive solutions must be found before we consume our way into another bad habit. We **need** to get it right this time as regards clean renewable energy sources. The first step in the correct direction is to do whatever it takes to make sure our alternative solutions are varied by technology, region, and fuel source. If the world used the sun and wind and geothermal heat to create heat/electricity -- all provided by different energy companies -- it would be a smart step in the right direction. We do a little bit of this today, but it's accidental. At one time, John D. Rockefeller controlled 98% of the entire world's oil reserves, refinery capacity, gasoline sales, and heavy fuels distribution, an unfortunate situation we still haven't recovered from. Ask yourself: what provides the power, the lift, the push, the strength, the raw materials, and the energy behind most things we use today. It's oil and the other fossil fuels. Oil, natural gas and coal provide over 70% of the electricity in the U.S. and 99% of the carbon emissions.

Now ask yourself: if just one person, or even five or six companies, controlled that pipeline, what incentive would it take for them to build an alternative? Viewed another way: If a benevolent being (the Earth) gave you a huge silo full of grain at no cost, you and yours would live well. Not only would the grain feed you, but you could also trade grain for other items you needed or wanted. Now, what if one of the facts you understood when accepting the gift of the grain silo was that when it runs empty you would *not be allowed to refill it* for one or two million years?

In that new scenario, most people would agree that the prudent course of action would be to use the silo's grain to create a different resource that would spew a different type of edible product. This new resource, by definition, would need to be up and running before the first silo runs dry! Farmers call this principle, *"Not eating the seed corn."* The human race has done nothing of the sort concerning oil. In fact, because of our lack of Market-Fences and a lax attitude toward anti-trust enforcement, we have devoured the seed corn. Furthermore, we let the source of this resource, the grain in this example, be controlled by people with monopolistic power, which guarantees that the prudent step of creating an alternative *will never be taken*. The people who run these companies know that as the oil runs out, they will actually make more money, even as wars break out over it. It's D-brain behavior, and it will happen.

Things We Know That Do Not Work

For U.S. readers: Sarbanes-Oxley won't solve the energy problem or any other problem for that matter. This law, along with another poorly conceived law, Bayh-Dole, needs to be completely scrapped. All that Sarbanes-Oxley has accomplished is to make the American economy and every player in it less efficient. Fighting wars with countries that have what remains of the world's oil supply won't solve the energy problem. Lowering oil taxes and/or drilling for more oil (see Mr. Hubbert) won't solve the problem. Another ridiculous and murderous option is to use government tax money to buy domestic loyalty in certain geographic areas with ethanol subsidies, which in turn drives up the cost of food. This increases the cost of food worldwide, increasing the number of humans dying from starvation.

Things We Can Assume Will Not Work If Tried

Attempting to predict which technologies will work by funding the pet projects Congress and the President currently favor is doomed to failure or extreme inefficiency at the least. This arrangement would put money to non-productive use, since the solutions would be sought from companies with huge economic clout -- the only kind that win these Congressional beauty contests. The F-22 Raptor is made from parts that come from 38 states. These are the same companies that typically throttle innovation and productivity breakthroughs. *The five things underlined in the above two paragraphs represent the sum total of the U.S. politicians' (Democratic and Republican) solutions to our energy problems for the last 50 years.*

As regards energy, we are cutting it close as a species. We've pushed our use of oil so hard and deep into Hubbert's curve, that the laws of supply and demand are creating enormous price and geopolitical volatility. Volatility for humans almost always ends in war. War is bad for the common man. Pressures on mankind from the lack of a suitable replacement for this dwindling resource, provided we don't innovate our way out of this mess, will starve/kill an additional two or three million people yearly, from now until a suitably efficient energy substitute is widely available. If the new energy needs described here are met in Asia through the burning of coal, the cost in human life and health will be the same in number, but from pollution.

A deep lack of understanding is the problem then. Our politicians habitually make the big problems *worse*—with their "solutions." Only through powerful anti-trust reform and Market-Fences will we harness the enormous power available by controlling the D-brain impulses of the business executives and properly motivating the scientists in these market segments.

The combined laws of thermodynamics are fundamental. They are the mathematical backbone behind the principle of supply and demand, **and** they cannot be cheated. Liberals and Conservatives alike (even President Richard Nixon imposed price controls) think they can legislate away the laws of nature. *They are always, always, always wrong.*

Scientists offer many reasons photovoltaics (one example of an alternative) will not work. These are red herrings. Because we don't have any anti-trust protections like Market-Fences that extend to and protect academia, scientists are for sale, making the denial and misinformation crisis even deeper. The multitude of forced disclosures by manufacturers concerning the suppression of negative test data on products and drugs known by the manufacturer to be harmful, but proven to be "safe" by their paid scientists and "lobbied" government oversight officials, speaks directly to this problem.

The Result Of Scientific Dishonesty

In medicine and energy in particular, scientific dishonesty is epidemic. The United States university system, the greatest university system ever built by the human race, is now subject to speech codes on one end and research grants tied to companies with vested interests on the other end. Every modestly cautious person should be deeply concerned about this. Humans will kill and lie everyday of the week for money (deer meat). To set up systems that do not respect this truth and are not designed to protect us from systemic human lying puts your family and our species at great risk by not protecting them adequately.

Liberals can never figure out why the American public loved Ronald Reagan so much. It was all D-brain. His "trust but verify" attitude made them feel safe. They liked a reasonable guy who was also willing to kill if necessary. But even Ronald Reagan, a very smart and intuitive man, did not understand the need for strong anti-trust laws. As a result, humans can now **own** genetic parts of other humans and parts of the human genome, all created by nature. Michael Crichton took this to the extreme in a form of modern satire with his book *Next,* but his points are all accurate, regrettably we lost Mr. Crichton this year to cancer.

The gap in the U.S. between modern medical science and law is creating a juggernaut of future problems. The above problems have fostered such highly unproductive behavior that it has already affected your standard of living. If not fixed, this will kill you one way or the other, either by ingestion of a drug that doesn't work or causes harm, or by economic deprivation as you struggle to pay the cost of a huge boomer population moving through their elder years in a medical system rapidly becoming less efficient (productive).

But I Digressed, Back To Energy...

Red herring #1 — The Toll of Line Loss. It will be necessary to cover cheap ground with photovoltaics (or any other technology), which by definition means remote, unproductive, and isolated territory. Since this moves production a great distance from where power is needed, the power and oil industry use line loss and insufficient infrastructure as a big excuse. But it is just an excuse used to fool the non-science-trained politicians. It is these same politicians that line up at the power and energy companies' doors to solicit "donations." The low temperature superconductivity of certain materials was discovered decades ago. It clearly demonstrated the possibility of constructing cost-effective power lines that don't suffer appreciable line loss. True, they won't be able to cross-large bodies of water, but that can be accommodated. These lines would be higher tech than the current aged and dilapidated transmission system, and would be dependent on cost-effective photovoltaic cells deployed along the lines to keep the superconductors cool. This is currently possible. I have also included in Appendix IV a *New York Times* piece on just how badly the current power line system needs to be overhauled.

Red herring #2 -- Who will pay for all this? In a healthy market, one with good anti-trust laws resulting in an ever-expanding economic cloud, the market itself would through the Market-Fence process. Currently we are not in an economically healthy market and don't have Market-Fences or even adequate anti-trust protections therefore I would advocate a 35% tax (which would increase by 1 percentage point every year for the first ten), on every gallon of gas sold. By *law,* that money could only be used for the R&D and eventual construction of renewable energy infrastructure. This tax would simultaneously slow down the consumption of oil and provide seed capital for a multitude of small companies that would be required to compete for this money with business plans that demonstrate productivity increases in *any* area of energy production except those using fossil fuels.

While I personally think taxes are the worst type of solution to most economic problems, we must face the truth about this particular problem: we've let it get out of control. We are so late in responding to the oil problem that additional taxation on that product is necessary and prudent. Humans alter their patterns very reluctantly. Large downward price swings in gasoline will drive up usage, despite what well meaning people *think* they would do with their consumption patterns -- if prices fall. A trade and cap system for the large carbon producers, <u>while useful if that's all we can get politically</u>, won't hurt *consumer* demand for fossil fuels, a big hole in that strategy. Of course excess carbon isn't actually responsible for global warming either, another big reason not to tie a very expensive cap and trade system to carbon reduction. Tying this huge system to an unproven indicator is foolishness.

A slowly rising tax on gasoline will keep the pressure on all of us to innovate, alter our lifestyle, up our productivity, and eventually force the substitution of an alternative fuel. I won't pretend such a tax would be politically easy. America's problems don't come from a lack of solutions, but from a warring C- and D-brain collective. A fuel tax doesn't discourage the creation of a finished human achievement; it discourages the use of a diminishing natural resource that human productivity didn't create. Best of all, the tax would disappear as the oil use did, a built-in sunset provision.

Red herring #3 – **A lack of clean, cheap silicon.** Although Bell Labs pulled a totally pure silicone wafer, the photovoltaic industry needs a *less* demanding material purity. However, for all practical purposes, the most efficient way to make a photovoltaic is to start at the higher purity silicone and then dope down to the desired composition. So, we need pure silicon. But, before Solarex in Frederick Maryland was BP Solar, they pioneered a technique for getting silicon pure enough for making photovoltaics using *woodchips as a purifying agent.*

This process is not being used commercially anywhere on the planet. Do you want to guess why? We humans have acted so stupidly -- would a chicken farmer allow a fox to guard the chickens? Why not? Because he knows the fox would eat every last chicken. Then the farmer, the egg company, the egg consumer, and eventually the fox would *all* starve. But those facts do not deter the fox. You cannot expect a D-brain driven animal, bringing home the deer meat, to develop a business plan and successfully implement it, if it endangers his personal supply of meat.

In large part, we are in the terrible energy crisis we face today because we allowed the oil companies to buy all the solar research companies, like the one in Frederick, Maryland that developed the processes for purifying silicon inexpensively. We should mandate that each solar company purchased by oil companies in the past split from the parent company and take 25% of the previous years oil company profits with them. The oil companies that either closed their solar company acquisitions, or never bought one, would have to contribute 25% of the previous year's profits to a fund that would be used to finance solar (or other "Green" alternative) start-ups.
Let an organization experienced in incubating start-up technologies, such as DARPA, pick multiple, promising finalists. Remember, we must have

multiple suppliers that compete for maximum marginal productivity increases if we want to quickly deploy alternatives with acceptable efficiencies.

We can make photovoltaic cells out of an abundant resource that can be cheaply purified. We can transmit the power to where we need it with virtually no energy loss. But, we have done neither because the patented knowledge is owned (and hoarded) by companies whose self-interest is to prevent the deployment of this technology. "How could something as boring as fossil fuels get me killed?" you may ask.

At a minimum, it is currently getting brave soldiers killed in Iraq and other spots around the world, and it's affecting your standard of living by decreasing your wealth. I won't say it's starving you to death, because I doubt people starving to death can afford to buy this book, but people are starving because of it -- you just aren't one of them…yet.

Harnessing D-brain drive with C-brain ingenuity would work to exploit each emerging technology in its own "Market-Fenced" economic cloud segment. **This is the *only* that scenario can save us.**

Book II, Chapter 5

Anti-trust Principles We Can All Live With, and Cannot Live Without.

> Rule 1. No corporation(s) and/or executives working for those corporations, nor any combination thereof, can ever own or control (through any mechanism), greater than a 40% share of any defined market. In *most* cases, this ceiling should be as low as 25%.

This must be an inviolate rule at the outset. Deciding which companies can be monopolies and which ones cannot, after they have become huge and influential, is a major reason the current rules are not enforced. Federal Judges see the inequity in the current system and simply don't enforce it.

If a company spontaneously grows into such a position that it controls over 40% of its market, it will by definition *have* to divest back down to the pre-determined limit. This will be seen as a violation by some potential monopolists of the "natural-born right" to win it all. To those concerned that such limits are somehow a fairness violation, I ask: Is it really beneficial for any one company or any one individual or any entity to "enjoy" this much economic success? In truth, the economic health of the controlling entity or person at this level of wealth is *ultimately dependent* upon the economic health of the consumer. As we have seen from previous examples, consumers' economic health is ultimately dependent upon not letting any person, company or single interest get a controlling level of market share. Therefore, divestiture is healthy for both the would-be monopolist *and* the consumer.

We need new laws and regulations to accomplish this. Some of the normal rules of anti-trust can still apply: e.g. all markets without significant barriers to entry would require no regulation at all. Big businesses have used multiple means, created by some of the cleverest attorneys in the world, to continuously run their companies in opposition to anti-trust principles for decades. They have preached "consolidation" and "efficiencies of scale," and then used it as a wrecking ball to destroy one competitor after another.

In the wake of their destruction are the scattered remains of what were initially vibrant and competitive companies, technologies, and segments of the economy. On Wall Street, they reward these "consolidations" with a higher share price. These share prices inevitably fall as the consolidated companies get bloated and inefficient while enjoying their unfair advantages. The bigger these companies get, the less able they are to compete or grow through marginal productivity increases *internationally or locally.*

Consolidation did not benefit the auto industry, the white goods industry, or the steel industry. I could go on and on. In fact, when viewed historically, only the deconstruction of the giant market killers has been proved to be successful in revitalizing failed monopolistic segments of the economy. Revitalization occurred only after they were forced to break up by bankruptcy or legislative fiat. This cycle is always true for an economy without pre-determined Market-Fences.

I have included for you a *New York Times* article on the Eastman Kodak Company. Although it wasn't the point of the article, please notice all the evidence against monopolies presented in this masterpiece. Not only didn't a monopoly benefit the consumer, as usual it didn't benefit the monopolist either. Please read with a special eye toward how, prior to its collapse, the Eastman Kodak management resisted any change in the business plan despite being faced with an obviously disruptive technology, the emergence of digital photography.

At Kodak, Some Old Things Are New Again

By CLAUDIA H. DEUTSCH

Published: May 2, 2008
ROCHESTER — Steven J. Sasson, an electrical engineer who invented the first digital camera at Eastman Kodak in the 1970s, remembers well management's dismay at his feat.

"My prototype was big as a toaster, but the technical people loved it," Mr. Sasson said. "But it was filmless photography, so management's reaction was, 'that's cute — but don't tell anyone about it.' "

Since then, of course, Kodak, which once considered itself the Bell Labs of chemistry, has embraced the digital world and the researchers who understand it.

"The shift in research focus has been just tremendous," said John D. Ward, a lecturer at the Rochester Institute of Technology who worked for Kodak for 20 years. Or, as Mr. Sasson put it, "Getting a digital idea accepted has sure gotten a lot easier."

Indeed, physicists, electrical engineers and all sorts of people who are more comfortable with binary code than molecules are wending their way up through Kodak's research labs. "When I joined, I knew my salary came from film sales," said Dr. Majid Rabbani, an electrical engineer who joined Kodak in 1983. "But I knew that I would eventually produce paychecks for others."

Kodak is by no means thriving. Digital products are nowhere near filling the profit vacuum left by evaporating sales of film. Its work force is about a fifth of the size it was two decades ago, and it continues to lose money. Its share price remains depressed.

But, finally, digital products are flowing from the labs. Kodak recently introduced a pocket-size television, which is selling in Japan for about $285. It has software that lets owners of multiplexes track what is showing on each screen. It has a tiny sensor small enough to fit into a cellphone, yet acute enough to capture images in low light.

Paradoxically, many of the new products are based on work Kodak began, but abandoned, years ago. The precursor technology to Stream, for example, pushed ink through a single nozzle. Stream has thousands of holes and uses a method called air deflection to separate drops of ink and control the speed and order in which they are deposited on a page.

Other digital technologies languished as well, said Bill Lloyd, the chief technology officer. "I've been here five years, and I'm still learning about all the things they already have," he said. "It seems Kodak had developed antibodies against anything that might compete with film."

It took what many analysts say was a near-death experience to change that. Kodak, a film titan in the 20th century, entered the next one in danger of being mowed down by the digital juggernaut. Electronics companies like Sony were siphoning away the photography market, while giants like Hewlett-Packard and Xerox had a lock on printers.

"This was a supertanker that came close to capsizing," said Timothy M. Ghriskey, chief investment officer at Solaris Asset Management, which long ago sold its Kodak shares.

Together, they have turned Kodak inside out. They exited a mainstay business, health imaging, and took the company back into inkjet printing. And they mined the patent archives for intellectual property, a step that is yielding well above $250 million a year in licensing fees. One recent example: Kodak is licensing out a method to embed a chemical signature in materials that enables manufacturers and

The Revelation

retailers to scan for counterfeit products.

"When it comes to intellectual property, they're finally acting like a for-profit corporation instead of a university," said Ulysses A. Yannas, a broker at Buckman, Buckman & Reid who has been buying Kodak shares.

And, perhaps most traumatic for a company that was known as the Great Yellow Father in Rochester, they eliminated jobs. Kodak, which employed 145,300 people 20 years ago, ended 2007 with 26,900 employees.

Analysts remain wary. "The stuff that comes out of the Kodak labs is impressive, but it does not give them a leg up on Hewlett or Xerox," said Shannon S. Cross, an analyst at Cross Research who rates Kodak a sell. Nor is she impressed with Kodak's consumer printer. "Consumers buy on the cost of hardware, not of total ownership," she said.

The Fujifilm Corporation, the Japanese company that was Kodak's main film rival, is not out of the picture, either. It recently moved its chemistry and electronics labs next door to each other. "If they work as a single team at the same location, R. & D. productivity is significantly enhanced," Shinpei Ikenoue, the head of research at Fujifilm, said in an e-mail message.

Still, analysts no longer predict Kodak's demise. "Kodak still has the most color specialists," Ms. Cross said.

There are a lot fewer of them, though. The research ranks have been cut in half, to about 1,000 people. "We watched a lot of chemists get downsized out of jobs," said Dr. Margaret J. Helber, an organic chemist who joined Kodak 18 years ago. "The rest of us soldiered on for several years, not knowing if we would remain relevant in the transformed Kodak."

They did, but in radically altered jobs. For one thing, researchers who rarely interacted are now expected to collaborate.

"This used to be a closed society, where some researchers kept their records in locked safes," said Dr. John D. Baloga, director of analytical science and a Kodak employee for 31 years. "Some of them were crushed when the secrecy went away."

Researchers also must now work with the business managers. Amit Singhal, a computer scientist who joined Kodak in 1998, said he had biweekly meetings with the business units. "I never used to see them at all," he said.

Indeed, until recently, functions like finance, marketing and research all reported up through their own hierarchies, ultimately to the chief executive. Today, everyone involved in creating, selling and servicing inkjet printers is grouped together, as are those dealing with cameras, sensors or other products.

"Finally, we have a structure that promotes commercialization of research," Mr. Faraci said.

The research chiefs do hold quarterly meetings to uncover technologies that can cross product boundaries. Marketing, operations and customer service chiefs meet regularly as well, to discuss which products and services can be bundled together for sale, or to see whether economies of scale can be achieved. But day to day, researchers and marketers deal more with each other than with their functional peers.

"Researchers invent something, demonstrate its feasibility, talk about commercializing it," said Julie Gerstenberger, vice president for external alliances. "But these days, it's all in collaboration with the business side."

Why then, may you ask, do big companies (and governments) always want to consolidate, legislate, finance, or steal their way into a monopoly? Can you imagine a U.S. Senator or Congressman from N.Y., voting to limit Kodak's market share during Kodak's heyday? Neither one would have understood how such a vote would have *saved constituent* jobs.

The executives at Kodak must have logically known that digital photography was inevitable, and that suppressing all innovation other than film was a huge mistake. But the executives couldn't help themselves because their D-brains were in charge and *wanted* a monopoly. Their C-brains just denied that they were doomed as a result, and ignored the camera *they invented.* Figuratively speaking, Kodak had tons of deer meat coming in, and was willing to alter everything in its environment to maintain that situation — regardless of how shortsighted its actions when viewed logically.

The D-brain is not logical; it is greedy. It's "Large and in charge." Kodak was no different from any other monopoly that has ever existed. From earlier chapters, we know why the D-brain behaves this way. It was trained, for hundreds of thousands of years, in an environment of subsistence hunting and gathering in which it *was* a net sum game, and where survival was enhanced or rewarded by taking everything one could. If the hunters brought home extra deer, their family would live longer. This *carefully evolved* Stone Age adaptation is completely dysfunctional in modern economies that obey the mathematical laws of thermal dynamics.

The D-brain favors:
On a personal level -- brute force,
And on a corporate level -- monopolistic behavior,
Both behaviors inhibit the processes which create more productive behavior.

In the end, it is only *productivity* that determines how much deer meat (wealth) is available to be distributed to all the needy humans – not how much control you have over an industry segment. Productivity is spurred by competition, which then leads to innovation, and technology advances. This is opposite to the goals of a monopoly.

For monopolies and oligarchies to be beneficial, the efficiencies created by economies of scale would have to outweigh the benefits created by ongoing innovation and productivity improvements -- **a ridiculous assumption**. Since monopolies have no strong incentive to improve productivity, these companies and institutions are highly detrimental to their economies in multiple ways, including the stifling of technological advancement and a lack of international market competitiveness. This has been demonstrated around the world *numerous* times.

Naturally, the senior managers of these super-sized corporations pay themselves a king's salary. CEOs of monopolies, and most especially oligarchies, are productivity/competition killers, and hence impoverish us all. Through their behavior, they cause untold human suffering through the creation of economic and societal crises such as: commodity shortages, recession/depression, unemployment, price gouging, even famine and epidemics. *Not one single exception to this observation can be found anywhere.* One major international corporation, Siemens, had a "cash" desk, a place where executives could go to get the cash they needed to bribe the officials necessary to influence their government contracts.

Over a few short years, by Siemens' own admission, they handed out tens of millions of dollars (at that desk) in bribes, with no paper trail or accountability whatsoever. It was the "grease" needed to keep the bloated government contracts they worked on.

How Do We Get Fooled Again and Again?

Currently, *de-regulation is one way that the* lies about "efficiency of scale" get presented to the population. And, we swallow it, over and over again, fearful to let jobs exist at the mercy of competition. De-regulation is presented as a way to create *fair* competition. In reality, the C- and D-brains of corporate management use the lack of oversight to create as many unfair advantages as possible in order to enhance their perceived survivability, which includes formation of a monopoly/oligarchy.

As explained before, our D-brains consider it too risky to compete fairly on a productivity and technology basis. De-regulation is not actual reform; it is merely a way of *selling* the chicken coop to the fox—with the accompanying illogical assumption that he won't eat the chickens because he now owns them. The government and the people, as well as the companies, all have a hand in this deception as a result of human evolution -- the D-brain likes *big*. Our D-brains like the idea of a monopoly. Our killer instincts admire a strangle hold on the competition. Corporate leaders cannot be trusted to change their desire or behavior on this topic despite an intellectual explanation of why it is a path to failure 100% of the time. They are only human. At our core, we are all D-brain hunters.

There is only one solution: Make companies and executives compete. Make it a corporate life-or-death struggle against many competitors without corruption or unfair external influence. This is the *only method* guaranteed to generate optimized productivity from us hunter/killers.

It isn't that competition is the best system we have to improve productivity and thus generate increased wealth for all; **it is the only system** our existing brain architecture will allow us to prosper under.

Current Regulatory Failure

Poorly-crafted laws designed to thwart this problem from a punitive view merely serve as a roadmap for the corporate criminals. As an example, Sarbanes-Oxley will never work, but it will get many people hurt through economic deprivation. I need not go on about the multiple failures created through poor governmental vision and influence selling, called bribes in other countries. This is so common that you the reader should already be familiar. Warning the fox (a company striving for the perceived security of becoming a monopoly) that society might jail him if he is caught eating the chickens (destroying the competition) will never work. Why?

Because of the inherent C-brain denial of risk, this approach has never worked and never will. The *only* solution to prevent economy killing monopolies is for the government to *pre-define* market segments with protective "fencing" through legislative action or federal mandate. Government must create clear definitions that the courts will be required to use. There can no longer be years-long debates concerning "What is the definition of our market" in civil anti-trust trials. This step is too costly and delays productivity improvements indefinitely. Currently, anti-trust trials are so costly that only big businesses can afford to bring anti-trust cases. Market Fences would obviate the need for businesses crushed by monopolies from seeking protection and redress in the courts. The prohibitive time and expense needed to get redress in the courts under today's laws is not lost on the monopolists. As currently constructed, anti-trust laws have perversely become the tool of the monopolists who are in violation of its core principles. Big business and their anti-trust defense lawyers may tell you differently, but in general, it is in their interest to lie to themselves, their customers, their adversaries, and their country.

Rule #2. No corporation nor executives working for those companies, through any mechanism, may own a controlling interest of a company in any market segment (as specifically predefined by the governmental agency in charge of Market-Fencing) that poses a conflict of interest based on three core principles: a threat to national security, a threat to food or energy production, and/or a threat to health and safety.

This rule is necessary because humans do not intuitively understand economics. We are naturally so greedy we don't care about whether we die from our greed next year—as long as we get paid more this Friday. Congress currently pays farmers to keep over 30 million acres out of food production in the U.S. -- this is an area the size of New York State. They do this to drive *up* the cost of food—one of the most unproductive ideas imaginable. You would think no one could possibly believe that this is a *good* idea.

If you feel this is a good idea, consider one important point; *economies only produce an increase in net wealth at the consumer level when prices are growing slower than wages.* Any action, which reverses this process, lowers everybody's wealth level. Many people are so ignorant on this subject they think we should pay more for things, to help the poor farmers, the poor car companies, etc. If you want to eat as much this week as you did last week, you must encourage laws and economic systems that obey the above productivity and anti-trust principles, not a misguided sense of "fairness." The C/D-brain-derived rules and inviolate physical laws discussed in this section of the book are not formulas or truths you can get around or change with legislation. The economic truths are like gravity. You could outlaw gravity and throw every Congressman off the top of a bridge, and they would all still go straight down!

Man-made laws change neither economic realities nor natural laws. Legislation is what people say; economics is the result of what people do. Everything humans do requires energy. Every output of human production requires energy. Everything we produce in goods and services, by definition, we must do in conformity with the combined laws of thermodynamics.

Is this starting to make sense to you? Would you like to see a really scary statistic as regards the United States? If you plot a graph starting in 1947 that shows the relationship of a price hike in oil and the U.S. economic status immediately following, you will find ten recessions over 60 years. Nine of the GDP contradictions (all except 1960, in which there was only a very small oil price increase) show the immediate relationship to the previous oil price increase. **Oil, in its role as our current primary energy source, is the most important productivity multiplier in our economy.** Here is what Larry G. Chorn, a PhD energy economist for the McGraw Hill Company, a publisher of trade journals, which focus on commodity products, has to say on this topic:

> "The United states built the largest economy in the world during the 19th and 20th centuries using domestic energy and mineral resources."

Those easy days now being over, he then points out candidly:

> "Consuming countries' failure to insure their energy security, may lead to declining growth rates, diminished standards of living, and growing transfer of wealth from importing to exporting countries."

Larry G. Chorn isn't a doomsday freak on the fringe of society; he is a PhD economist for the country's leading energy trade publication. Again, Dr.

Chorn states:

> "There is a definite relationship between a country's gross demographic product (GDP) per capita and its energy consumption per capita, as rising consumption relates to rising GDP. A simple lesson for governments: make more energy available and the economy will grow."

Isn't that what we have discovered through the Combined Laws of Thermodynamics explanation of the economy? It is surprising that these points still need explanation. Dr. Chorn observes later in his article that because some national companies asserted rights over domestic oil fields, new discoveries have slowed dramatically:

> "Countries are aggressively trying to tie up oil reserves and production capacity to sustain their economies using national oil companies as an extension of foreign policy. As we have seen, oil consumption is intimately related to GDP and, therefore, growth, geopolitical influence, and ultimately a national population's well being."

His conclusion was entitled, <u>Observations</u>. (OECD stands for Organization for Economic Cooperation & Development and BRIC stands for Brazil, Russia, India, and China.)

> Observations:
> "A scenario in which OECD and BRIC countries are in competition for energy security is clearly under way. Sustaining economic growth and maintaining standards of living are already front and center in the actions of several countries. National oil companies, implicitly and explicitly are being used as diplomatic agents in this scenario. However, as the US and Canada do not seem to have agents in the 21st century reincarnation of "The Great Game", they continue to rely on the multinational presence of US and Canadian based oil and gas companies to ensure supply to offset domestic shortfalls."

Dr. Chorn overlooked the enormous help the U.S. government has given U.S. oil companies over time, including war. That aside, his major point is that as countries industrialize there is much more new competition for oil supplies. There are not enough oil reserves in existence for all of them. Dr. Chorn thinks oil companies are the wrong instruments to get that oil. Dr. Chorn is right. Our governments and their corporate oil partners have served us very poorly for the last 20 years as regards energy security. Essentially, governments left consumers in the middle of a stream without a paddle, no batteries in the flashlight, and kids to feed sitting on the rock next to them. And guess what? Only effective, strong, well-written anti-trust and Market-Fence laws could have prevented this. This is also the only way out.

The myriad of solutions necessary to prevent or cure a problem this big can't blossom in an environment sterilized by monopoly and oligarchy. I could put literally hundreds of examples in the book to demonstrate this point but I'll put in just one. For decades now, the power and energy companies have owned most of the solar and wind turbine technologies; and, as a result, very little progress has been made toward commercializing the technology.

Just five years ago, GE, not a direct energy supply company, bought up the wind turbine unit from Enron Corporation for only $300 million. Last year, (2007) GE announced 10% EBITDA on *more than $4.5 billion* in wind-turbine sales.

Even if you could magically wage a successful war to control all the remaining world's oil reserves, you would only buy a few more decades of energy while killing tens of millions of humans in the process. Truly that would be the most "losing" of all propositions, allowing our Greed and Fear to decimate love. We need to let anti-trust do its job, so that properly motivated engineers and executives at companies like GE can save us all.

Book II, Chapter 6

What If Something Really Bad, Happens?

What if the "shit really hits the fan?" It can. Oil is not the only example.

Epidemics are the most likely near-term crisis for several reasons. They have a very high frequency (every 33 years or so); they target exploding populations (humans); and they can evolve, strike, and mutate very rapidly, making treatment and a successful immunization plan difficult. The sentence above was written before the H1N1 virus gained national and international fame but this recent event only points out how vulnerable we are. Our approach for handling this type of problem is wholly inadequate at the moment. Humans don't have the political systems, the information systems, the medical systems, food or medicine reserves or even enough **trust** from country to country to fight a problem like this.

We cannot rule out severe climate change, natural disasters such as Earthquakes, asteroid hits, and volcanoes, or a worldwide economic depression triggering war and/or a fuel/food crisis. I lean toward large-scale economic disruption and/or epidemic as the next big long-term threats, because those are historically the most likely problems. Global warming is almost certainly a red herring, although global *cooling* is a genuine concern, according to the Revelation.

The Cooling Image

The image in the Revelation raising my concern over cooling looks like a circle, covered by a white mist, a mist I am *sure* is water vapor or cloud cover. It covers more of the circle than at all other times. When the image is fast-forwarded through time, the mist (clouds) partially clear out of the first circle -- the one below is also mostly white, reflecting back so much solar radiation that it doesn't warm the planet properly. If the first circle is the atmosphere and the second circle is the planet's surface, then you can see how this would work. Water vapor and dust, not CO2, are the culprits here. No matter the means by which you get it into the atmosphere, whether by volcano (under water or on land) or by asteroid strike (ocean strike or land strike), almost certainly (other images demonstrate this) *it is not exclusively by a too warm environment*. Environments warmer than ours is currently, have existed on Earth before, for long periods of time, without creating the cloud cover that radiates too much solar radiation into space, which is *the* problem according to the Revelation. This vapor/cloud phenomenon is what initially causes the cooling, which then causes those clouds to precipitate snow and ice. When the clouds clear (this is the first circle), the second circle now radiates too much heat back into space, *keeping* the planet cool. CO_2 is actually one of the planet's tools *to fight* this problem. This is one of the areas mentioned for which we will need geo-engineering to combat/manipulate our environment. If we do not, again, partial de-population will occur.

Asteroids and Comets

An asteroid capable of wiping out our species, and most other species, will hit the Earth. *That statement is a certainty* provided we don't do a few simple things to prevent this occurrence. There is an excellent lecture on this subject given by Ivan Bekey, the former Director of NASA's Future and Advanced Programs for 18 years, detailing the math much more beautifully than I am able.

Here is my shortened version. We **are not** surveying all the sky for large orbiting objects that pose a potential threat to Earth; and we need to be. We don't have a governmental or non-governmental agency responsible for designing and launching a space-based interceptor if we are lucky enough to identify a threatening object in time. These are steps we must take *in advance* of the discovery of a comet or asteroid bearing down on our planet if our response is to be effective in time to save us all. The Jet Propulsion Laboratory (JPL) team has the proven ability to hit such an object, alter its orbit, and save us all... but only if given enough time to launch the interceptor while the object is still far enough away that only a small course or speed alteration is necessary. Believe it or not, we could survey all the applicable areas of space for only a tiny fraction--one billion dollars, or 185th of the money the U.S. government spent in 2008 to save an insurance company called AIG. If you ask me, we bought the wrong insurance policy as a nation.

In summary, the United States needs to unilaterally establish a fully funded space-object monitoring program, design and build interceptors, and begin working on a legal framework for launching such an interceptor provided we find a target we need to hit. (Waiting for the UN to figure this out will make a devastating asteroid strike a certainty.) Large objects that cause regional or planetary catastrophe hit approximately every 26 million years. Very large asteroids that hit land throw up huge amounts of dirt and dust into the atmosphere, obscuring the sun, and killing plant life and the species dependent upon that life. Life in the deep oceans and some tiny land-based life forms survive, but everything else dies. This is exactly what extinguished the dinosaurs, and they were amazingly long-lived, tough competitors compared to humans. The last very large asteroid strike, 65 million years ago, killed 90% of the species on the planet. Smaller asteroids of about a quarter mile in diameter create regional devastation but not global annihilation. This may sound like good odds for those of us alive today... but read on.

Apophis 99942

Discovered as a threat in 2004, there is an 800- to 1,000-foot-long asteroid (Apophis 99942) hurtling toward Earth at 40,000 miles an hour that could extinguish much life on Earth, were it to impact us. It will actually have two chances of striking us: once in 2029, and a second time in 2036. In 2029, Apophis is calculated to pass *between* the Earth and the moon. In fact, it will come inside the orbits of most of our communication satellites. The following quotation is from cooperative research by NASA/JPL, Caltech, and the Arecibo Observatory. They describe the results of their radar observations of this potentially hazardous asteroid along with an in-depth analysis of its motion. I quote them directly:

> "The future for Apophis on Friday, April 13 of 2029 includes an approach to Earth no closer than 29,470 km (18,300 miles, or 5.6 Earth radii from the center, or 4.6 Earth-radii from the surface) over the mid-Atlantic, appearing to the naked eye as a moderately bright point of light moving rapidly across the sky. Depending on its mechanical nature, it could experience shape or spin-state alteration due to tidal forces caused by Earth's gravity field. This is within the distance of Earth's geosynchronous satellites. However, because Apophis will pass interior to the positions of these satellites at closest approach, in a plane inclined at 40 degrees to the Earth's equator and passing outside the equatorial geosynchronous zone when crossing the equatorial plane, it does not threaten the satellites in that heavily populated region."

They describe a fast-moving massive object 18,000 miles from Earth's surface! Most of the calculations done concerning this asteroid have a combined margin of error larger than the margin of safety. Assuming the math is sound and it does miss in 2029, Apophis has 1% chance (according to JPL) of passing through a "keyhole" area of our gravitational field 600 meters wide that is projected to line it up for a direct hit upon its return in 2036.

Not to worry the reader further, but I found the most disturbing part of JPL's presentation to be the admission that they only know 50% of the parameters necessary to make a final calculation concerning the impact probability for this 2036 pass. Combine this with the fact that we are currently only surveying a small section of the relevant areas of space known to contain the asteroids and comets as the direct result of our lack of will to provide funding, and you can see how ignorant we can behave because of denial. This denial of risk will cost us our life as a species-- if not from Apophis, then from another yet-to-be-discovered comet or asteroid. If you get a chance to hear Mr. Bekey lecture, I highly recommend it. He is an experienced astrophysicist, and quite a gentleman. Perhaps that is why our governments aren't listening.

What You Should Do – In General

In any type of disaster, your primary mission is to stay safe, protect your immediate family, and wait for the cavalry. Hopefully, in 30 to 90 days, the government will restore order and services. Your first obligation is to your immediate family. This is a D-brain mandate.

This is also exactly the type of group the government would want you to form to survive. Small groups, family-based, where people know each other and a natural chain of command already exists, make for hardy survival teams. Forget about government help in a catastrophic emergency; their first priority will not be you.

So, you will need:
    a. Water
    b. Dry food
    c. Weapon & ammo
    d. Hand crank powered radio
    e. Bandages & antibiotic soap/cream
    f. And most of all, a plan.

Believe it or not, a plan is the most important thing…and it costs nothing. The plan should be a realistic summary of what you would do, and what you expect others in your immediate family to do, in case of an emergency.

Here is a crude sample:

1. Gather family at home.
2. Barricade home. (If necessary, could you do it—do you have the supplies?)
3. Ration supplies until order restored
4. Protect home if needed.

Even a simple list like this makes our complete lack of planning for such emergencies very obvious. Most people do not have a plan to gather their family, much less any barricading supplies, water, or dried food. There are two or three weapons per household in the U.S., so plenty of those are available. The vast majority of Americans (more than 95%) have *never made any* emergency preparations. Let's suppose hypothetically that you heard about a potential disaster on the radio. Do you really believe that thirty minutes later you would be able to purchase and store emergency water and food? After the H1N1 virus was made public in 2009, many sporting goods stores in the U.S. had to limit ammunition supplies *because of overwhelming demand.*

If you want to be very careful about survival, you could build a log cabin on the other side of the mountain from the major city. Install a back-up power generator and big fuel tanks, along with solar heating, a garden plot, and plenty of grain storage. Then, you and yours could survive a year without the restoration of order. I don't think this is practical for most people, but it is statistically a good idea. This plan is called the Ark principle, for self-evident reasons. The long-term goal of the Ark plan would be to build a family compound where the following ideas get incorporated: A house or set of buildings designed for the nuclear relatives and their children to live out their lives together with other family members, if possible.

The compound should have solar, biomass, geo-thermal, or some combination of the above, which would provide the majority of heating and cooling needed. This was thought to be optional for many builders over the last few hundred years but they will all be proven wrong. Eventually every house will have this in one way or the other, whether or not we want to pay for it.

The compound should be stocked with rations and water (well water is better if possible) for the house's expected occupants plus a factor of 100 %. You also need weapons sufficient for a short loss of civil order. All members living there will be familiar with, and have assigned duties in an emergency.
There you have it: One Ark. Different Arks could link up and share plans and tips via the Internet.

I'm not going to start a web site for this but if you want to base one upon these ideas, I encourage it. Contests for acquiring the most calories in the longest possible storage mediums, at the lowest cost per calorie, would be a good example of things for people to discuss on such a site. The goal here would to make a simplified Ark solution for the average family, spend less than $500 total, and be implemented throughout 50 percent of the population. Complete Arks with family compounds will be rare, but I suspect this idea will catch on one day, if mankind survives long enough.

Book II, Chapter 7

Privacy Law And Lawyers

The train wreck between the C- & D-brain on abortion created many negative societal consequences. In particular, it has hurt the privacy issue at a time when we should be strengthening privacy protections as a result of technologic advances, which make it simple and cheap for businesses and government to spy on millions of people.

This violation of our privacy may not sound dangerous or connected to the other topics in this book; however, you will see the connection clearly drawn in the next few pages. The conflict created by falsely stating "privacy" concerns as the legal basis for making abortion rights untouchable has the potential to impact the shared-power arrangement between the U.S. government and the U.S. population. By linking the emotionally-charged C/D brain issue of abortion to the unrelated but extremely important issue of personal privacy, we have preemptively destroyed our opportunities to protect our personal privacy over the years.

The Supreme Court, Congress, and Presidents treat any issue concerning privacy and privacy protections like a hot potato in the modern-day U.S. This is a very damaging state of affairs during a technological revolution that has made privacy protection one of the most important issues of our times. Unfortunately, because of our focus on Roe v. Wade and the war on terrorists, the advancement of privacy rights is frozen.

Most Americans have no idea how much data about their daily lives are available to corporations and government—household habits, driving habits, spending habits, reading habits, traveling habits, virtually all the things that define who you are, what you do, and how you do it. Moreover, these two monolithic entities *don't care about you as an individual or citizen.* There is no point in attaining the peace and happiness of finally understanding your mate if you both end up in a gulag or concentration camp at the end of the day.

Your behavioral data get used in millions of ways everyday, and has the potential to impact you negatively. This gathering of information about you, performed in the name of serving you better or protecting you from enemies, tramples your basic liberties as a citizen as well as your right not to be bothered by those wasting your time and money. How often after receiving an inferior product or service, have you paid the bill anyway because the offending company threatened to put a bad mark against your credit? Millions do it every year in the U.S. alone. This occurs because a company can report negative information to one of the 3 major credit bureaus, impacting your life negatively and dramatically, without any of the necessary consumer protections that are morally and logically demanded by this arrangement.

Who are the people reporting/recording your data? What law gave them the authority to do what they do? Why were they allowed this power? I never authorized *anyone* to set up a database compiling credit data about me—did you? Isn't that information mine? A woman has the right to keep a pregnancy private but I'm not allowed privacy over whom I bought tires from, whom I bank with, or how much credit I have? How upside down is that logically? How can the recording of potentially negative (credit) misinformation be possible without any review mechanism? Why is this allowed without a challenge process that works? Ever try to erase a mistake off your credit report? The credit bureaus are allowed to do this…because they just did it. They never asked anyone if they could. They just take your data, *which represents your life activities,* from the corporations with which you interact. These corporations never asked you if they could share your data with anyone. But, they do. They use this data to tell "interested parties" where you are, what you eat, in what stores you shop, and there's nothing you can do about it under current law. They *charge others* for details about you and your life.

There is a badly defective notification process, which by law requires notification of the consumer upon whom they are recording an unpaid debt. But, this "protection" is required from the company that wants to share negative accusations about you with the world, and may have started the incident by charging you for something that is fraudulent and for which you really shouldn't have had to pay.

> Corporations do not police themselves adequately enough to be given this power over consumer data.

The entire credit rating/collection process excludes an independent source of analysis. The current process assumes that the original company that provided the poor service or product will honestly comply with the notification process. Even if their claim against you is challenged, it won't prevent credit bureaus from recording erroneous information about you. This is a terribly unbalanced situation - *and criminals, corporations, and banks have figured out how to use it against you, and to their advantage.* When dishonest companies or common crooks lie, cheat, and don't provide the service promised, they simply sell the bad "receivable," your purported obligation to pay them, to a less-reputable company known as a collection agency. In this manner, "reputable" companies use collection agencies to get around legal requirements that are supposed to protect you. You have no recourse other than the official process of wrangling with the credit rating companies in an effort to change your credit record. Usually this happens after you've been hit with a downgrade in your credit score. The correction of mistakes happens slowly and very infrequently.

Criminals, as well as Fortune 500 companies, use similar business strategies and techniques in this segment of the economy. If the product or service is knowingly defective, they still collect your money -- or you will have bad credit. To hold the companies or criminals responsible in court would cost legal fees far in excess of just paying the false charges. This is called consumer "green mail." When the criminals cannot improve upon the techniques of licensed corporations, you know our business culture has reached a new low.

Let us suppose you hire a contractor to work on your home. He signs a contract with a rental company to use equipment needed to perform the job. This is a very common practice. The contractor requires an advance payment, and you comply. The contractor shows up to perform the service. During the job, you notice he is failing to do something necessary, or is doing something vital in an incorrect manner. You discuss the problem with him, and he promises to fix it. After much discussion and time, the problem still isn't fixed. So, you ask him to stop work. He demands payment. When you refuse, he leaves. Not only does he not finish the job, he processes another payment on your credit card.

This is not obvious fraud in the credit card company's eyes, so they *will process* this transaction. In fact some banks process millions of dollars in fraudulent payments, in almost identical circumstances, to the *same offending* corporation. How many transactions? How many are challenged? Sorry, that info is private, but you the consumer have all your information exposed.

A month after fighting with the credit card agency and the credit bureau to make sure this doesn't impact your credit negatively; you get a bill from the equipment rental company. The equipment the contractor rented for three weeks has a $4,000 bill due. They want the money from you. You laugh and say, "This has nothing to do with me. I just fought a big battle with the credit card company over this guy; he's a crook who didn't finish the work. Plus, I never rented anything from you."

Imagine your surprise when the equipment contractor informs you, that because the "work" was performed on your house, it doesn't matter to them

The Revelation

whether it was finished, done correctly, or that you didn't sign the rental agreement. The truth is that he can file a mechanic's lien against your house, which will be automatically recorded until you legally challenge it, because that is where the work was performed! Don't believe me? Ask a lawyer.

Your blood turns ice cold. Your Dad is in the hospital and you expect to close on a home equity line in the next 60 days to help with that expense. A lien will derail that loan, or force the payment of the rental fee before the loan is granted. You plead with the rental company representative. He says it is company policy and hangs up. Why?

The rental company doesn't care because your fraudulent contractor rents from him *every week*.

The divorce was a bad one. The ex-wife hated you because she found out you were sleeping with another female. Her D-brain made decisions that she denies to this day, that cost tens of thousands of dollars in extra legal expenses for no concrete gain. They were all things her lawyer actually encouraged her to do because he knew it would run up his bill. He also knew that she would agree because her emotions were in charge. She falsely reports you to the state child support agency for non-payment. They automatically record a non-payment of child support against you—without a review *or even* notification. After you find out and challenge this false claim with proof of payment, the state credits you for the payment, *but cannot or will not force the credit agencies to change the record marring your credit.*

After you spend months dealing with a credit repair agency that you have to pay, the unpaid notice on your record is changed, but the entire event is not completely expunged and never will be. Credit bureaus simply will not do this. You live with this mark against you the rest of your life, a violation of your rights in so many ways that you either learn to accept it or it will make you go crazy.

Sadly, when the events like this happen at emotional points in time, many people do go crazy. They get violent and even start killing (mostly men choose this action). The male D-brain programming and patterning understands a mate and offspring; but it cannot even begin to understand lawyers. Lawyers essentially represent the C-brain run amok. They represent and enforce intellectually decided laws and customs without regard to the D-brain needs and preservation of the individual or family. The legal profession is an axe splitting the D-brain away from the human that needs both brains to be happy and survive. The D-brain cannot respond verbally, and so emotions and violent reactions are to be expected. Lawyers in over-abundance magnify this problem. Companies have privacy because they use lawyers for their internal planning and operations. What lawyers do for them is considered "privileged." Most U.S. courts consider this privilege of legal privacy absolute.

On the other hand, since you do not run your personal life by consultation with a lawyer, nothing you do is covered by privilege. The courts, through judges, insist on seeing your complete trail of private documents, even when their own rules say they shouldn't be looking at it. Lawyers understand this game and play it maximally. They can level one untrue charge after another against you, routinely lie to the judges, and have no fear of retribution since you cannot charge them with slander or libel for statements made in the written lawsuit or in the courtroom. Why? The answer is privilege, of course.

To make matters worse, the U.S. courts consider the attorney and client *one* for legal purposes, so you can't even charge them with conspiracy. You simply have no recourse but to spend money to clear the lawsuit and move on. This reminds me of a boxing match where lawyers and their clients can punch you, but you cannot punch back, except by spending more money on the system and profession that brought the (false) claims against you in the first place. In Britain, these outrages forced the adoption of a "loser pays" system, which has reduced frivolous lawsuits in that country enormously. With some modification, this system would work very well in the U.S. It needs to be instituted as soon as possible.

Want different examples of legal abuse? Look no further than the unnamed law firm in Washington D.C. representing the government of Serbia to manage a contact-and-influence campaign with the ultimate goal of blocking the independence of Kosovo. This firm does this work despite the fact that the independence of Kosovo is a stated goal of the law firm's government and your representatives. U.S. soldiers were fighting in the Balkans to stop ethnic cleansings and ensure the security of Kosovo.

If you, as an American citizen, did something such as this law firm, contrary to the policies or laws of the U.S. government, you would be investigated and likely charged with a crime. Your activity would at a minimum get you "watched" and questioned. But when a lawyer does it (as long as they register their activities with foreign governments), they aren't even censured. In fact, they will make hundreds of thousands of dollars in fees. These particular lawyers aren't any more corrupt than the others operating today. Most lawyers in the U.S. would represent Satan if he had enough money. Actually, for Satan, they would extend credit. Most lawyers could always stand to have the devil owe them a favor.

I don't want to spend more time on a subject on which much has been written. But, if you want a better understanding of the enormity of this frightening problem, read: *The Rule of Lawyers: How the New Litigation Elite Threatens America's Rule of Law*, by Walter Olson. I particularly like this book because Mr. Olson also points out how lawyers, up to a certain point, encourage productivity by correcting bad behavior. But after that point, defined as a percentage of the population, lawyers become leeches, sucking productivity out of the system with expensive and unnecessary actions. The U.S. is so far past the break-even point on that issue that we all suffer dramatically. Lawyers shamelessly use class action litigation, emotional distress, divorce and personal injury, as well as lobbying, to represent any point of view, even ones openly harmful to the rest of society, as long as they think they have the potential to make money.

Lawyers know a loser-pays system would empower the poor and the weak -- the current prey in our system, people that simply cannot pay enough for lawyers. By correcting the current imbalance between the poor and those with money, who currently don't fear the legal system, lawyers would only lose paying clients. So, the United States Trial Lawyers Association (in all its various forms, they're like the old KGB, who also kept changing its name) has fought this proposal for years. It hasn't been much of a fight.

The U.S. government system is completely rigged in the trial lawyers' favor. Practicing lawyers are part of the political professional fraternity with a functional majority of our elected representatives. Even with a medical care and drug cost crisis looming, tort reform that would limit medical lawsuits and their associated costs is "dead in the water" in Congress. The elected representatives simply get too much money from lawyers to upset their campaign cash flow with critically needed legislation.

It is shameful that at some point, years ago, lawyers and their disingenuous solutions began to choke productivity, making us all poorer. For those of you married to one of these sharks, I am sorry. If you're in love with one, at least make him/her work pro bono in Africa, in a health clinic, for a year before you'll agree to tie the knot—test the love.

Book II, Chapter 8

Japan, Germany, etc., and World Unification
...Also the Anti-Christ

Many people actually believe there are worldwide organizations (e.g. Trilateral Commission), which have a secret plan to unite the world and strip people of their national sovereignty and civil rights. There are such organized secret international groups, but their purpose is more mundane. Their usual goals are to enrich their members financially or through increased power status.

Furthermore, some people believe these worldwide secret societies, and organizations, are headed by Jews, the U.S. government, or any other group that makes them feel threatened. This would include groups that speak a different language, are a different color, or are wealthier. Many conservative Christians believe there is a biblical prophecy stating that the first man to organize a successful effort for world unity will be the Anti-Christ. Talk about a difficult hurdle for a visionary and loving prophet to overcome! Think what conflict that prophecy creates. In the eyes of over one billion Christians around the globe, the first person to seriously attempt worldwide human reconciliation, peace, and a basis for global cooperation is *automatically* demonized a supernatural devil bent on the destruction of all mankind. Only creatures as full of denial and hatred as humans could have set up a trap, this diabolical, self-defeating, and idiotic. If and when some nice guy does try to end war, poverty, and other worldly troubles through a sort of global unification structure, he'll probably get shot for his efforts.

The book you are now reading is solely designed to give you information— not to fight straw men, create new ones, or start any political movements. Even where I use very political examples like in the last chapter, they are still just examples. It is the brain process that is important. That is what I want you to recognize. If you come to recognize *the process*, you will be able to make up your own mind on all the politics.

Use your C-brain and your D-brain. Be aware of C-brain denial and D-brain greed and emotional motivations. Use your head when exposed to all the nonsense people say and write. Become educated as to facts. Then, judge for yourself if what you are being told makes sense. This is how Winston Churchill understood Adolf Hitler before everyone else. Churchill knew that to understand Hitler he would need to read what Hitler wrote, listen to his speeches and watch his actions—nothing special or magical. This same process will work for anyone.

Our ultimate choice as a species is rather simple and straightforward, though the path may be difficult. Either humans will find a way to cooperate, unite, and work together to solve our problems, like the world's food and energy production shortage, or we will fail to solve these problems and instead join into groups that have common interests; through war and starvation, we will proceed to kill those who will not cooperate, have something we want, or have committed some past, perceived, sin. The choice is ours to make; at stake is the survival of our species.

In other words, humans are doomed to kill each other in greater numbers, both intentionally and unintentionally, until they work out some type of cooperative world government. No other viable choice exists.

I am not referring to anything resembling the United Nations. The UN is based on C-brain Collectivist thought and does not take into account the D-brain basis of man's behavior. Consequently, the UN has a zero chance of success. The U.S. Federal model is a much better example. However, to be practical on a world stage, it will have to be modified so the balance of power is tilted more toward the individual nations (States) and less toward the Global Unified Government (Federal) side than currently exists in the U.S. In fact, it would be better for the U.S. to rebalance the split as well, but that is not the point here.

Such a Global Unified Body (GUB) could allow wide regional diversity including language, while lowering trade barriers, tariffs, commerce restrictions, and, most importantly, currency and civil rights discrepancies. Ironically, the GUB would have little to do with diplomacy, particularly as we define the word today.

In Truth, Diplomacy is a Waste Of Time

When people make war on other humans, the country's leaders are rarely influenced by diplomacy, because their first allegiance is to someone/something else. In the tyrant's case, it is his own self-interest. In the case of the elected leader, it is his electorate. All the U.N. hopping from foot to foot in speeches and proclamations about Darfur hasn't saved one life.

The people in charge on the ground in Darfur (there are many) are all trying to kill each other—death to their fellow countrymen is actually the point. There is no desire for international aid by those in power, unless the aid can be stolen to feed their own troops. Then, of course, the food is allowed to land. It is only when the person you are negotiating with fears or respects you, that diplomacy can accomplish something. This almost never happens, but it was worth mentioning. The above reasons are why the Chinese and the South Africans both use diplomacy differently than we do. The South Africans and the Chinese make the same point about President Mugabe of Zimbabwe. Mainly they feel you must let native populations work out their own issues—right up to election rigging, the killing of political opposition leaders and supporters, and beyond--to mass genocide against some of their own ethnic minorities.

This is the *expected* Revelation answer from regimes such as the Chinese, which practice very little power sharing with their own people. From a D-brain perspective, what do you realistically expect Communist Chinese leaders, leaders that rolled over their own democracy protestors with tanks, to do about rigged elections in Zimbabwe? Especially Zimbabwe, a country that sells China important raw materials below market value.

U.S. participation in the fraud that is the U.N. has real consequences for those impacted populations. What did you do – as an American -- to save those people? "We sponsored a really tough but fair and balanced U.N. resolution coupled with economic sanctions, all of which we knew had no chance of success from the moment we conceived them. But we've done something!" In reality, covering up "No action" with "Impotent action" is much worse than doing nothing and just saying, "Sorry, I can't impact this, you need to fix it yourself."

In the U.S., we came to our power-sharing arrangement between our government and our people in steps. We fought two wars with our mother country for separation and liberation. And then we fought an internal war, bloodier than all the rest, over our enslavement, transportation and genocide of our black minority, who began their time here in America as a slave labor force. Suffrage was a recent event, ladies and gentlemen.

The Civil War in particular nearly destroyed the U.S., but was also a baptism through which much of our strength is derived. Every country must experience some version of this. The more complicated the population mix, generally the bloodier the resolution process. Using this scale, a Pentagon General should have predicted a long, bloody process for the unification of Iraq under an occupying army. Notice the parallel between the steps of a BHU becoming a CHU: a mandatory stepwise process is the only permanent way to change/evolve, both personally and nationally.

In the end, it was war that changed Iraq, not diplomacy. People who hate war are going to be offended by this line of reasoning. I am. But the implications of the C- and D-brain arrangement are inescapable. Humans are killers and they generally don't settle anything of importance with permanence unless they kill a lot of fellow humans to settle the point. This exemplifies why diplomacy actually does more harm than good. Create a system with accountability, and people will be accountable. Attempt to create one without risk, without consequences, and you'll choke productivity or start a war every time.

The attempt to create systems that yield only positive results is a C-brain collectivist trait that humans must learn to stop repeating.

Did you know that DuPont scientists were looking for a "perfect" solution, a chemical propellant that would **not** interact with any other substances in our ecosystem when they stumbled upon chlorofluorocarbons (CFC's)?

> " Eureka!!!" They must have thought at DuPont!
>
> "This stuff doesn't react with anything! Every single atom of this stuff we ever make will just sit around in our atmosphere forever and do nothing at all."

They actually thought that was a good idea. They were *purposely* seeking an inert substance. Aside from the fact that they were wrong and that UV breaks chlorofluorocarbons down into, among other things, Chlorine, the yellow gas that then dissolves the ozone layer, the bigger question remains, what if those guys *had been right*? If CFC's hadn't broken down, wouldn't they have eventually changed the gas composition of our atmosphere and contaminated our breathing supply with all those CFCs themselves?

Mathematically then, the error that DuPont scientists committed when they invented and manufactured CFCs is the same error the SEC lawyers committed by allowing the unregulated fabrication, use, and leveraging (to 10X the world's GDP) of derivatives. In both cases, by attempting to mitigate *all risk*, the entire system becomes contaminated, and the entire gas cloud that represents that system is cooled simultaneously. In the DuPont case, it literally diminished the ozone cloud rather than cool it.

Attention to this balancing of risk highlights the fundamental misunderstanding *of the positive role of risk* in our world economy. This principle must be developed, understood, and incorporated into the creation of our economic and political processes in order to increase productivity enough to satisfy our D-brains and make armed conflict less appealing. In other words, war can be minimized, but not without recognizing that humans are killers, highly competitive animals, and that our economic and legal systems need to account for this – rather than deny it.

> We can never honestly outlaw war. What we **can** do is to remove as many of the underlying pressures that cause it as possible. The brains of humans worldwide, sensing those changes, will be less likely to permit/desire war. Remarkably, it is a gas cloud of economic pressures, societal behaviors, and individual brain programming, multiplied millions of times, which create war. That's why it can't be outlawed. You can't change a snowstorm by outlawing it either. Both are the result of many other "gas cloud" factors that human laws do not impact.

As regards C-brain priorities: there need to be civil rights protections applying to the freedom to worship or not worship, and the freedom to select leaders and vote upon laws (shared-power). For this to work well, the majority of people in these societies must be C-brain aware (educated), and D-brain wealthy (productive). This is precisely why representative power-sharing forms of government imposed by the U.S. and/or other well-meaning countries upon an educationally unprepared and marginally productive population **always fail**. BHUs don't fare well with policies designed to work in CHU heavy economic systems/society. However, therein lies a great opportunity with the Japanese, the British, the Italians, the Germans, the Australians, and all the other similarly educated and technologically advanced countries.

> The U.S. should begin working toward a world federalism model with these and similar countries by creating a worldwide Bill of Rights, a worldwide currency, a worldwide trade organization (WTO is a pretty *good* start), a dismantling of the worldwide defense systems (e.g. NATO), and a worldwide framework for the division of power between the overarching federal type system and the more powerful individual member nations.

A rotating Capital should be done. Nations at war would immediately lose membership until war had ceased. Membership in the Federation would be the carrot side of the equation to reduce war, since using a stick to eliminate other sticks is already the existing definition of war.

Why Disband NATO?

Mutual defense treaties are a bad idea. They increase the likelihood of a regional conflict, because of core brain principles recorded elsewhere in this book. The first is the C-brain and D-brain balance. In the European Union (EU) for example, the existence of NATO created a "safety" umbrella for the industrialized nations in the EU to live under. This "safety" umbrella allowed the EU members to decrease defense spending to only one or two percent of GDP, on average, while one partner in the mutual defense network, the U.S., made up for this deficiency by continuously spending five or six percent of their much-larger GDP on defense.

Being protected (versus self-protecting) changes the brains of the protected population, altering their C & D-brain balance and making them more vulnerable to extinction. The survival side (D-brain) component of the citizens of a country defines the survivability of that country as a whole. The supported countries' lack of responsibility for their self-defense shifts the burden to one member of the alliance. This is likely to cause an economic imbalance, which eventually will be another cause for war. Populations that are truly responsible for their own defense, both in blood and money, exercise that responsibility more rationally.

Democracies who bear the cost of war rarely attack, (nor are they attacked by) other democracies. There is an even more dangerous side to the imbalance created by NATO. When the D-brain side of a population (country) is weakened, the C-brain becomes too powerful. *With unchecked authority, the C-brain creates non-competitive and non-productive social and political systems doomed to failure, all in the name of Fairness.*

We have witnessed various social welfare states produced in Europe during the 70s, 80s, and 90s crash in economic and social chaos. The pain is not over for them yet, because the cause is still present! The insidious nature of excessive C-brain power and denial not tempered by D-brain fear, results in the creation of self-destructive social policies that inevitably lead to economic weakness, as evidenced in low GDP growth rates.

Social welfare states create debt to continue servicing the population's needs. Eventually, decreased productivity and social redistribution leads to the rationing of resources among the population. This is disruptive to human social order, and ultimately leads to war. Just like you cannot subsidize a BHU to become a CHU (it is something the person must evolve through), you cannot defend another group of people for an extended period of time, either militarily or economically, without changing their brains in a way that is bad for the protected group *and the* protectors.

Foreign Policy

Regardless of personal beliefs, this is a very cogent argument for not having a purely professional military, or "Volunteer Army" as we call it in the U.S. today, but rather require that all citizens spend two years in military service between high school and college. This would allow for very important and necessary C-brain and D-brain development, concerning the concept of self-defense, in each human serving. These massive brain-evolving institutions can have a profound long-term effect of the success of the species. An organism that doesn't intimately understand its own defense mechanism is flirting with extinction.

This two-year draft system definitely would benefit the U.S. as well as all the other countries the U.S. currently defends through treaty or sphere of influence. We previously discussed Market-Fences as a way to keep each market's economic playing field balanced, competitive, and segregated from unequal forces. The "Fence" prevents a problem in one area or segment from infecting other market segments worldwide. **This** *same principle* **applies to foreign affairs.** In other words, instead of the U.S. government's ridiculous effort to contain Communism, which led directly to our involvement in two costly wars, we should have been focused on *catastrophe* containment, or what has now become known as Resilience Planning. A better C/D-brain understanding would have pointed to the high probability of failure behind much of the U.S. foreign policy over the last 50 years. This approach would have highlighted the probability of economic stagnation from such policies, and the importance of letting those BHU heavy countries evolve on their own to a more productive economy, *even if it was under the "Tyrant of the day."*

Power-sharing populations like the U.S. are ultimately never at peace with the non-power sharing countries in which they have selected the leadership. Our experience with placing the Shah of Iran in power demonstrated this. Consider the residual animosity in the Persian population toward the U.S. as another good reason for not imposing our C-brain values, policies, and tyrant preferences on foreign populations. It is the indigenous population of the dominated country, not the folks back home in the U.S., who must tolerate daily life with the tyrants our government chooses for them. This should serve as guidance for the world's powerful countries as regards future actions in the Middle East, Africa and Asia. Put more simply: What **power-sharing** (U.S.) population would ever be happy with the practices, politics, and leadership of a tyrant their government backed or installed in a foreign country, no matter how benevolent the tyrant? And much more importantly: What **non-power sharing** (Iran in this case) population will ultimately be happy with the practices, politics and leadership of a tyrant (Shah) that another country picked or installed -- no matter how benevolent?

Suppose (not so hypothetically) that the U.S. had supported twenty different tinhorn dictators throughout the world over the last 40 years, ostensibly to guard against the scourge of Communism. Then suppose that over time, the populations in those counties became increasingly C-brain aware, in large part through the U.S. support of education and technology, and eventually wanted to power-share with their own government. This is something the Revelation describes as inevitable. This unavoidable situation perversely puts the U.S. population squarely at odds with the new power-sharing, democratic movement! And even more importantly—the power-sharing populations in those countries, our *ideological* allies would then be at odds with the U.S. population.

Introducing: "The Swinging Door"

This is why the "Swinging Door" principle explained below, needs to be part of the GUB Federation philosophy. If the countries in the GUB Federation have their own armies which the individual populations pay for with their own money and blood, **and** the individual nations can leave the union easily (e.g. to war on a neighbor), they will do this knowing that the responsibility, cost, consequences and loss of blood has been brought upon themselves through their own actions, or the actions of a domestic tyrant. This means shortened survival for externally-placed tyrants. Iraq's Saddam Hussein is the perfect example of a tyrant that lasted so long because the U.S. funded him, and then ultimately had to pay dearly to remove and execute him.

Fencing-out regional disputes from the rest of the world would allow them to resolve on their own, without disrupting the money and flow of goods to the other nations. I am sure this philosophy will drive the peaceniks, the UN, and the C-brain-heavy collectivists crazy. As counter-intuitive as this concept appears, it would serve to make war much less likely. And when war does occur, it will be shorter in duration. It would involve fewer countries and lead to less loss of life and treasure. War will not be completely prevented, but every population will clearly know what they are sacrificing in terms of personal resources and economic benefit when they leave the Federation to make war. And, they will know that their war has to end before they can join the GUB again. Making the expulsion and re-admittance easy to do will reduce the occurrence and duration of war.

The U.S. "Elasticity" Problem

Unfortunately, while a wonderfully-crafted document, the U.S. Constitution is a poor model for a worldwide federation because of a drafting problem. It purposefully channels the power for the development of future laws *away* from the individual states. For this reason, if it were to serve as a GUB worldwide model, it would fail to serve the rights and needs of the countries that belong to the Federation. Unfortunately for the U.S., the result of this design flaw has been the constant seepage of power into the Federal system. It cannot be reversed under the current document. This is highly destructive to the health of the individual states.

The "elasticity" clause, or "necessary and proper" clause as some people have labeled it, is the root of this growing problem in the U.S. constitution. This needs to be amended to respect the rights of individual states to define and execute optimum solutions to some "necessary and proper" problems that might conflict with alternative solutions that work best in other areas of the country. This would be doubly true of the GUB as it relates to the nation-state members of the union. The member nations would obviously need much more legal and social independence, compared to the U.S. Federal model, as well as police powers, since these powers are cultural and relativistic. Additionally, the member states will retain all military powers for their country.

The most important requirement for the GUB to work is a core group of civil and religious liberties that are protected with bedrock hardness--a major concession each national government must make in trade-off for membership. Some type of standard or test would have to be agreed upon to test the member states on their continued good-faith efforts as regards these principles. If people can trust the courts and government with their individual rights, democracy flourishes, where they can't, it will not last, the D-brain is happier with tyranny.

We Need A "Swinging Door" or How Lincoln's Love Saved US

This brings us to the "Swinging Door" principle. This concept must be incorporated into a GUB constitution though it conflicts with virtually all previous international cooperation models. Since each member state will be responsible for its own country's defense, the proper function of the swinging door component would make dismantling NATO and other mutual defense treaties a necessity. As explained above, this will lessen the likelihood of war. Mutual defense treaties serve only to line up powerful countries on two sides of international issues, making it more likely to lead to armed conflict rather than prevent one.

Early on, Lincoln tragically spent a lot of time and energy on the idea that once a state was ratified, it did not have the right to leave the Union. This doesn't make democratic sense, nor can it work in practice. It wasn't Lincoln's insistence on this concept that allowed the Union Army to win the Civil War. Instead, it was Lincoln's change in thoughts about slavery, and the resultant issuance of the Emancipation Proclamation, that *created* the United States as a union— for the first time.

Civil liberties are not the gift of democracy; they are the ropes that keep it together. Civil liberties are the only foundation upon which democracy can truly function in name and spirit. As we have seen, "level playing field" competition in a market will reward you with extra productivity and wealth. Preserve civil rights among a herd of humans, and they will reward you with Democracy, and a vastly increased competitive production leading to increased wealth distribution and societal happiness.

To prioritize, let's *not* look at our choices from an American point of view, where we elect our leaders, participate in the creation of our laws, *and* have civil liberties. Where would you be happier?

> In a country where you can participate in elections every few years but **do not** have civil liberties?

OR

> Where you cannot elect the nation's leader but **do have** significant civil liberties?

Look around the world and examine where each of these situations exist. In every case, it's better to be in a country where the people have civil liberties regardless of the leader. *Most* tyrants routinely stage elections. Look at the real-world example of this in Iraq, yesterday and today. Saddam Hussein had elections, but they were rigged. The ropes of religious and clan loyalties are much stronger than those of democracy. Hence democracy there today is still just a word, even though someone else now rigs the elections. Democracy is a fashion, the imposition of an occupier, a curiosity, but not a reality worth sacrificing for -- like family, religious sect, and clan. The things to die for in this type of religious and clan-based country are highly valued by the D-brain...because it is a strongly D-brain-dominant society. In a generation, with education and modernization, Iraq will transform on its own to a more C-brain-aware society. At that point what its citizens will value, and defend to the death, will change, and democracy will flourish.

Iraq will continue to prove this theory in black and white. If the Sunni majority government protects the civil liberties of the Shiite minority, that country **will not dissolve** into endless civil war. If the Sunni power structure attacks or fails to protect the civil rights of the Shiite, the Shiite will wage war inside their country on their own countrymen. Much of the fighting in Iraq has *really* been about this issue, not the American troop presence.

Global Unified Body and Civil Rights

These three things: an unyielding GUB bill of civil rights, dominant state's power on most issues, and a "Swinging Door" principle for joining or leaving the GUB must all be present for a world federation to work.

Under civil rights, the following core rights must be protected:

1. The right for all citizens: of reasonable arrest only, including a speedy and fair trial with legal representation for all defendants. Trials and hearings to be conducted by unbiased juries and judges.

2. The freedom to worship a God(s), or not, free of interference from others.

3. The right to individual privacy, both electronic and physical. The standard for this should be modeled on the privacy a human had the right to expect *before* today's electronic enhancements.

4. The freedom to speak freely, including reasonable opposition speech to the government.

5. The right of peaceful free assembly.

6. The rights of a free press.

7. The right to separate themselves politically from the (GUB) union via a quick and fair process, similar to the process to enter the union, The Swinging Door principle. Nation states, which cannot uphold the civil rights protections to the GUB standard, or are at war, will be expelled swiftly from the union.

8. The right of protection from: the passionate majority, government abuse of power, and violent, persistent, or pervasive discrimination based on race, religious or political belief.

The Revelation

How the Atomic Bomb Saved Millions

Prior to WWII, the U.S. cut off fuel and scrap metal shipments to Japan after the Japanese began invading the countries (China, Korea) around them. *The Japanese considered these actions part of an undeclared state of war, which based on the communication trail between Secretary of State Hull and the Japanese seems in retrospect to be an **appropriate interpretation**.* (There is significant truth to the theory that Roosevelt understood the Japanese, waiting and baiting their D-brains into attacking the U.S.) The Japanese responded by attacking Pearl Harbor. We responded by attacking the Japanese and the Germans. In the end, the U.S. developed the atomic bomb and dropped it on Japan. Roosevelt knew that to enter the war any later could mean the loss of England and that meant invading fortress Europe from Greenland or Virginia, neither of which he considered plausible. The problem was, one day prior to Pearl Harbor, the nation was still divided over the issue.

The dropping of the atomic bomb, inarguably, saved *Japanese* and American lives. This fact is *rarely* brought into debates over the morality of dropping the bomb. Here are the numbers presented in a way most people have probably never seen them. Most discussions about the decision to drop an atomic bomb on two Japanese cities involve two points: the estimate of *American* lives saved by dropping the bomb, and the theory that a demonstration of the bomb's power on an unoccupied target would have been sufficient to persuade the Japanese High Command to surrender.

The problem with this discussion is that Truman didn't order his advisors to present him with **all** the casualty numbers, just the American casualties of a conventional war with Japan. Their estimates were based upon a beach type landing assault on the home islands, after devastating artillery bombardments from the sea and air. The Japanese had little air or sea power remaining at this time. Since the American loss of life was expected to be approximately 200,000-400,000 men, the Japanese loss of life would have been approximately *ten* times (conservative estimate) that number. Those 200,000 American servicemen were going to die while killing over two to four million Japanese. The fire bombing raids already conducted in the six months leading up to wars end was estimated to have killed between 400,000 and 800,000 Japanese. And they were just in the *softening* stage; this was the lead up to the invasion. Strictly from a numbers point of view, the atomic bombs saved those 200,000 American lives *and* at least another 1.6 million Japanese lives (after subtracting the approximate 400,000 human losses at Hiroshima and Nagasaki). To a mathematical certainty, we can state that the atomic bombs saved at least 1.8 million more human lives compared to the traditional sea and air campaign planned against Japan.

As regards discussion topic number two: How about just a demonstration of atomic power to the Japanese Military High Command? Think about that from their D-brain point of view, and you instantly see the problem with that expectation. Even after the first bomb was dropped on August 6th, the Japanese High Command refused to surrender. It took another bomb and nearly 10 days more before they admitted defeat. Do you really think the Japanese would have surrendered having been given a "demonstration?" The demonstration idea is pure C-brain denial at its most extreme. Read *Imperial Tragedy*. This amazing book, written by one of Japan's military elite from that time *for a Japanese audience*, describes in great detail exactly what the military officers and leaders of Japan knew about atomic energy and its potential power. The answer was almost nothing. There was only one Geiger counter in the entire country of Japan before the bombing of Hiroshima.

The Japanese would not have surrendered because they would not have understood or believed what we were showing them. The U.S. only had two bombs available at the time. A failed demonstration was not an option. Those dropped atom bombs, though an awful commentary on the nature of man actually saved many more Japanese lives than any other possible alternative. *As bizarre as it may sound to your C-brain, the dropping of the Atomic Bombs on Japan to end World War II was the correct and loving choice. It was the compassionate choice.*

The Power of Productive Cooperation Over Poverty & Destruction

Look what happened in Japan post-World War II. With few natural resources, a bombed-out industrial base, and a worn and hungry population, both Germany and Japan rapidly used nothing but productivity increases, fueled by trade and externally imposed managed market segments – the first try at Market Fences!-- to create more marginal wealth per capita in a shorter period of time than had been ever created before by humans. Historians mistakenly point to the Marshall plan as the explanation. I am not saying it was unimportant, as it was an act of compassion/love that serves as the true key to the universe. However, the level of wealth invested by that plan was microscopic compared to Japan and Germany's GDP growth numbers. Furthermore, Marshall Plan-type investments have been tried elsewhere around the world but failed to promote productivity and wealth. Why? Japan and Germany produced astronomical returns because the population was a mix of C-brain-literate (therefore with potentially high productivity multipliers) and D-brain terrified people. With just small amounts of seed capital and highly managed market segments, the framework of the economy was uniquely suited to promote massive marginal increases in productivity that literally created great wealth out of thin air. That is the engine that grew into the Japan and Germany of today. So, are these lessons being applied around the world? In China, they are. Why are these lessons not being applied everywhere? I think, by now, you know the answers.

In Conclusion...

Political unity for the human species under some type of loose, worldwide federation has become imperative. For reasons mentioned (and some unmentioned) in this book, it **must** be done *within the next 20 years*. This is needed to fight the threats of global hunger, poverty, disease epidemics, ocean depletion, asteroids, global warming/cooling, environmental changes, war, economic disasters, and resource depletion -- all leading to inevitable human conflicts. If we choose to ignore this imperative, our species will suffer multiple and severe depopulation diebacks. I believe there are computer models that predict this depopulation scenario if many things don't change. These events will actually look and feel worse than a quick extinction, as we will fight, struggle, and kill each other every time a worldwide crisis occurs.

Let's unite enough to avoid this scenario; call it the Peace Machine Doctrine if you like. The idea is that we need to unite because our collective human footprint on the Earth's ecosphere has gotten too large. We must either act as stewards of the planet's resources, which takes coordination, or exhaust them one by one, as we race toward extinction, killing the oceans, the land, and the air -- and finally, ourselves. Not to sound melodramatic, but I do consider that last sentence an accurate statement scientifically. There really is no other choice than cooperation if we wish to survive as a species in an environment this limited, at such great population numbers.

Finally, look at this from an evolutionary point of view. If cooperation among humans is the evolutionary adaptation that gave us superiority over all the other species on the planet, which I assert is the fact, what is our certain destiny if we fail in that adaptation?

Book II, Chapter 9

About WAR

"War. Good God!! What is it good for? *Absolutely nothing!* Say it again!" This is a catchy little tune from my past, but unfortunately, the truth is much more complex. Let's assume for a moment a "simple" two-nation war.

> Since by definition most individuals correctly consider self-defense an absolute right, this makes war, in at least 50% of situations, good for something. This point is valid no matter what else is assumed.

If attacked, the fear of death and the D-Brain reflexive fight for survival, cannot be diminished/extinguished in ours or any other predator species without leading to the *extinguishing* of the species. All predator species without a reflexive defensive strategy share the ignominy of extinction. This fact makes keeping the necessary D-brain defensive reactions and competitive drive alive and well for species survival. Since we can no longer rely just on primitive family structure and/or a subsistence environment to provide this, we must artificially create social structure for it.

> To accept the importance of integrating these D-brain principles into our social structure, we must first admit how important war was to mankind. It was the ultimate expression of his competitive nature, and contributed enormously to his survival during the subsistence environment period, which was *most* of the time man has roamed the Earth.

So if self-defense makes 50% of all war justified, and if we can't survive without the self-defense urge, what then about the other half of the story—those that attacked first? Are we to assume, out of distaste for war, the attackers are always wrong? Are political, religious, economic, and nationalistic goals, ever an adequate justification for attacking another group of people with weaponry? This is an academic question because it involves C-brain rationalizations for war--which is always a D-brain-driven activity. Therefore, discussing C-brain rationalizations for war is a waste of time.

The Revelation

Certain populations develop a hatred for other populations. Based upon what we've learned about the D-brain to date, it should be obvious how and why this happens. This emotional D-brain mistrust of others can easily be manipulated into popular support for a war among a significant enough percentage of the population to initiate a war action. In non-power sharing populations, those populations need not even be consulted or support a war action, but have to fight anyway.

Still, the question asked above is important. In many situations, such as trade embargoes of necessary or essential products, or financial actions that threaten a nation's economy, you could argue a military response is itself a *necessary* form of self-defense. This was Japan's argument for the attack on Pearl Harbor, even though in this particular case the U.S. shut off oil and scrap metal shipments *after* the Japanese attacked China and Korea. Even if some excuses are worthy in just as little as 10% of non-defense war actions, it still makes war in the majority of cases justifiable by our core brain programming, programming that is necessary for species survival. War today is, nonetheless, incredibly wasteful.

War has gotten considerably more civilized and humane in recent times, with modern armies *mostly* attacking other modern armies with the goal of avoiding civilian or collateral damage. Still, it wasn't too long ago that Sherman burned the city of Atlanta to the ground and Osama bin Laden targeted civilians with his jetliner bombs.

In countries where the D-brain power structure, typified by tyranny, is still dominant, their armies *do attack and kill civilians and will continue to do so*. In some cases these armies are primarily constituted for that purpose, not to repel foreign invaders. Ninety percent of the new Iraqi army's time, money, and energy will be directed domestically for decades. This will change when the majority of the population and its power-sharing structure become dominantly C-brain driven.

Using our C-brain to assign a logical goal to war is fruitless and damaging, because people then talk about how to create solutions to fix the phantom issues instead of the true issues underlying the conflict. Sherman was burning his fellow countrymen's city *at the Civil War's end* out of anger and hate. *Ostensibly* Osama bin Laden's main gripe with the U.S. was the stationing of 5,000 service personnel at the Prince Sultan air base in Saudi Arabia. Those

U.S. service personnel were there at the request of the King of Saudi Arabia. It is not possible to enter Saudi Arabia without being invited by the Saudi Government/King. (The King *is* the law.)

Bin Laden himself took money from the very Saudi leaders that had requested the "infidels" to encamp at the Prince Sultan Air Force base, and used this money to fund his attack on U.S. civilians. Bin Laden was taking money from these and other Arab leaders right up to the day of the 2001 World Trade center attack, a neat trick we could call D-brain denial. The leaders who gave bin Laden the money to attack their protector nation (U.S. forces were guarding Saudi oil fields and borders) were exercising C-brain denial. None of this is logical; it is the C/D-brain run amuck. Why do the Junta leaders in Burma kill unarmed peasants and Buddhist monks? None of these actions is logically justifiable either. But they make sense if man is a predator, a killer, a hunter that enjoys not just the spoils of the hunt, but the competition and the kill most of all.

Holding onto power, exercising power, and killing others, are the only things tyrants really care about. Religious power is also deemed power worth killing over. Osama bin Laden's real gripe is the Western world's refusal to recognize and share any of his power as bestowed upon him by his military accomplishments, his religion, his family status, and his cultural heritage. He fought along side Mullah Omar and the rebel Afghanis to free Afghanistan from the Russians, with U.S. and Saudi assistance. When the U.S. denied *either of them the right to rule their prize*, a blood feud began that will not end until both these men are dead, the West is overthrown -- or the Caliphate orders forgiveness. Radical Muslims consider Arab lands sacred, meaning no foreign military personnel can be allowed to step, sleep, or speak blasphemy on those grounds. To some, *any* word by a foreign fighter is considered blasphemy.

This is an impossible standard and will lead either to the destruction of these target populations *or the killing or changing* of those with this particular philosophical outlook, such as radical Islamic warriors. Bin Laden wants to be a complete religious tyrant in an age when the world is opening up -- at a time when different cultures are trying to connect their C-brains without bloodshed, war, bondage, slavery, and genocide. His counter-flowing tide of destructive thinking is the expected, natural, uncontrolled D-brain response to the larger movement of C-brains connecting.

To this day, the Taliban throws acid in the faces of Afghani girls who attend elementary school, or who are incorrectly attired. Any philosophical or religious interpretation that culminates with its followers throwing acid in the faces of young children cannot be *the truth*. Not the true philosophy of God.

Not the true philosophy of love. Not even the true philosophy of survival. The Extreme Muslim Fundamentalist Movement (EMFM) will inevitably die out from the weight of its own untenable ideas and aspirations. The absurd alternative would be the complete destruction of the all non-Muslim human populations here on Earth – an unlikely event.

This C-brain triumph over D-brain inequities was as true for the Southern plantation slaves as it will be for the Saudi housewife, though most people do not see how these two are connected. The same C-brain tide that swept away slavery in the South will eventually get that woman a driver's license in Saudi Arabia. Ironically, this is true for the very reason Osama bin Laden believes it will not happen: because, as we will prove later in the book, God is Love. His tool here on Earth, a cooperative human species, is still developing.

This does open another interesting question. Who "rightfully" possesses any piece of ground or water? I see this debate held *ad nauseam* in one publication after another, throughout the world. It is usually connected with war. The entire debate as framed by modern societies is pure C-brain denial and lies. Ancestral rights, native populations, and religious rights are all rationalizations that mean nothing. Other than outright agreed-upon land purchases, the above justifications are mere fabrications meant to justify the force used to acquire land from someone else.

The only system man has ever used for the possession of land is:

Whoever has the physical power to possess, control, or influence control over a particular piece of land or body of water, "owns," however temporarily, that land or water.

Americans, whose C-brains do not agree with the "Might is right system" described above, are using this principle currently by removing people (Mexicans) from the U.S. who have an "ancestral" right to the land we now control through might. No need to discuss what we did to the Indians. It was genocide. Despite our desire for "fairness," no war reparations, or apologies, nor any other C-Brain nonsense will work. Solutions such as those *contribute* to the hate cycle I described earlier. Usually the minorities who are intended to benefit from reparations get hurt, often before reparations are even paid.

Think about the 300,000 Berlin housewives, husbands dead, buildings leveled, being raped, almost to the last woman, by the occupying Russian soldiers in 1945. These Russian soldiers were bent on venting their hatred of

Germans. These same Hausfrau's supported the war with the French and English over WWI reparations. This new war then got their country bombed flat, practically everyone they loved killed, and themselves brutally raped by Russian soldiers. Hatred breeds hatred.

Hate breeds hate, but God is Love. One must choose to accept forgiveness in order to accept love. And, one must choose to forgive in order to give love.

France and England wanted so badly for Germany to pay for WWI that they insisted on war reparations, crippling the defeated country's economy and limiting the essential supplies available to the German population. Two decades later, the Nazis took just a few weeks to go around the Maginot line and devastate France in a loss of life and economic damage worse than France's share of the war reparations could have ever been worth, *if* they had been collectable.

Even those who ostensibly support détente and cooperation between neighbors don't fully believe this diplomatic C-brain principle because it conflicts with a D-brain truth, the desire for more deer meat. The Democratic Party in the U.S. would collapse without union support, and the unions want illegal immigrants in this country (they work cheaply) thrown out. They want trade with countries that have low wage labor restricted. So, the Liberal Democrats look past their principles and are as anxious to build a wall between Mexico and Texas as the archconservatives. *Such a wall is a very, very bad idea for everybody.*

Finally there is the question of war spoils—usually territory. The fight to remove people who have squatted on a piece of dirt that supposedly belongs to the non-squatter is a very old fight. No one can say for certain which group of people possessed which piece of land first except for *two* basic things we know to be true. The first is that our particular genetic ancestors came out of Africa and killed all the other species of humanoids roaming the Earth. They then spread out from there, taking by force everything they could. Technically speaking, if we want to open the "Ancestral Land Rights" argument, we theoretically should start by giving the Africans the rest of the world. But then again, we are all Africans since we all descend/originated from one human ancestor 160,000 years ago. The previous two sentences highlight the absurdity of "Ancestral Land Rights" as currently viewed by human law and customs such as 'Birth Right Tours," in Israel.

Secondly, since all living humans descended from a common ancestor less than two hundred thousand years ago, we are, although we express our genes in a variety of colors and shapes, ancestrally speaking, all the same race. Therefore, we all have the same rights to all the land. We really aren't Jews,

Blacks, Italians, Arabs, Asians, and Indians from each and every continent—we're all humans from a single parent about 160,000 years ago. This human "sameness" also leads back to the truth of my definition—Might is right when it comes to borders. To suggest anything else is to waste time, money, and blood on pointless negotiations. Land negotiations only benefit the power brokers, never the people.

The UN, which has as a core purpose the settlement of land and border disputes, is a useless entity. We can observe this statement's accuracy if the UN's actions over the last 30 years are examined. Unfortunately, like welfare, the UN is not an experiment without a price. If you appoint an impotent, dummy agency to police the world, then the world's effective military forces don't enter into any conflict unless they choose to do so for their own self-interest. This results in the U.S. using its military for Gulf War I in the defense of a country we needed for oil, and precious little effort spent for the dead and dying in Darfur, Cambodia, Burma, etc. – people that we do not need to maintain our standard of living.

The single most surefire way to predict whether a conflict will continue without resolution is the presence of UN peacekeepers, or the imposition of UN sanctions and/or a UN sponsored peace plan. Ask yourself one question: if your life was at stake in an area "protected" by this organization, would you flee or stay? Refugee statistics from areas the UN is trying to police and stabilize would support my assertion that when millions of D-brain-dominant people get to vote with their feet, they run from the **lack of safety** those little blue U.N. helmets provide. For all of the above to be understandable—we only have to agree that humans are killers and liars, a fact pointed out a hundred different ways throughout this book. Once you accept who we really are, it's easier to see the truth about how and why humans behave as they do. Humans give peace a chance as often as their D-brains think they can afford to—which is almost never.

There is a way to prevent getting involved in a string of constant wars; and, that way is reasonable deterrence.

Reasonable deterrence is **not** 60,000 warheads like the U.S. and Russia had aimed at each other at the height of the Cold War. That was economically very wasteful. I don't think General Curtis LeMay's decision to fly bombers along the Russian air space borders day and night with live nuclear loads, for years and years, was necessary, though he *definitely removed all doubt* in the Kremlin about what might happen if they got careless or stupid and preemptively attacked. Even in a vodka-induced haze, Kremlin leaders understood what would happen if General LeMay ever got the "Go" word from the U.S. President. He would have turned those bombers inland vaporizing every Russian city in order of descending population numbers—without a second's hesitation. He was a D-brain-heavy man, feared and understood by D-brain Kremlin leaders, as well as the Japanese High Command--that he bragged would convict him of war crimes if the U.S. had lost the war in the Pacific.

Deterrence is like a gun in William Raspberry's home. It's OK to be a bleeding-heart Liberal, if when push comes to shove, you're ready to drop all pretenses and kill the threat. You don't need to spend a fortune to do the job either; just signal your willingness, capability, and intentions to defend yourself. That's all that's needed 99% of the time. I say 99% of the time because the other 1% is much trickier. I simply don't see a rational way to deter the Hitler-type nationalistic D-brain maniacs, or the Osama bin Laden-type religious D-brain fanatics that believe, at their core, that the killing of civilian men, women, and children makes God happy. This is precisely the problem the Israelis have with Hamas and the radical Palestinians. Competing groups must be willing to admit each has a right to exist before they can accommodate each other as neighbors. Otherwise, you are forced to fight to the death. No room at all for the Peace Machine. Imminent death is the only currency another D-brain killer understands. The war continues until they don't want to fight anymore, or are dead.

Osama bin Laden is right about one thing; there is only one God. He is also right that the God he speaks of is interested in *his* actions. But until those actions are motivated by love, Osama has no chance of success. He is doomed to fail because God's plan does not include the purposeful killing of the innocent, the weak, women, children, and civilians when Osama can protect his family *without* doing these things. I don't dispute that Muhammad's holy teachings include a warning against allowing infidel soldiers to encamp on holy ground. But would Muhammad endorse an act of war on innocents, women, and children, to chase these devils away - especially at the certain cost of thousands of Arab lives? The soldiers at Prince Sultan were only an offense against dignity, not Arab life or limb. Osama's response is unholy. Saladin and Muhammad would not have poisoned their own wells to kill infidel soldiers, and then filled the water skins of their own people before leaving for exile.

> War itself must have self-defense or survival (not suicide) as its core purpose or it is doomed to failure.

Every great warrior learns this lesson from history books or at the point of a sword. Too many American Presidents are equally ignorant on this point. Fortunately, the majority of regular people are not ignorant on this point. The general populace won't support wars, that their D-brains are not comfortable with—wars that are truly lacking the element of self-defense. The citizens' lack of resolve in this situation is always death to the war effort. I call the recognition of this phenomenon the "Vietnam Lesson," which, amazingly, is *still lost* on the American people and the Pentagon.

Politicians and Pentagon planners argue that the strategy of having only limited deterrence is suicide in today's high-tech world. They falsely reason that because asymmetrical attacks can kill so many, *a huge defense bureaucracy must guard against these attacks.* This is D-brain lying coupled with C-brain and D-brain greed. Such a deterrence strategy is extremely expensive. Since the entire economy is an interconnected gas cloud, the dollars spent on 60,000 warheads instead of the 2,500 we now consider an acceptable level of deterrent, did cost us many lives. The more money you spend not feeding, clothing, and sheltering your people, the fewer people there are, and the less wealthy we are as a country. Today's war on terror is yesterday's Cold War: a big price tag associated with little actual security.

Suppose the government could prevent *every* terrorist attack by spending an extra 100 billion dollars every year. Most people would support this policy—for a while. But as no attacks ever materialized, the public would eventually want that money back to spend on other priorities. Even a number of small collateral damage type attacks would not change this.

The proof of this lies in the fact that al Qaeda *was attacking* the U.S. during Clinton's presidency, and the American people just did not notice or care because the death toll and locations of the attacks were limited. Look at the *decline* of the defense budgets under Clinton, the dates of the U.S.S. Cole bombing, the two African embassy bombings, and a few other skirmishes, like the one that occurred in Mogadishu. **We were under attack by al Qaeda then...and just didn't notice or care.**

The American people and Congress blame the CIA for being surprised by the New York Twin Towers attack, but they need to share the blame. What should the CIA have done after the Cole bombing and the embassy bombings in Africa, all of which occurred with complete public knowledge printed in major newspapers? Should the CIA have taken out a full-page ad in the *New York Times* declaring Osama bin Laden was at war with us? You could watch Osama on the Internet declaring he was at war with the United States years before the 9-11-2001 attacks occurred. That wasn't special limited intelligence; it was *common knowledge* that the American people didn't absorb or respond to -- because we didn't care. We were in C-brain denial until it enraged our D-brains on September 11.

To divert the U.S. population from its constant economic focus, people need to see a lot of dead Americans. This will never change because the D-brain's priority is the hunt for deer meat (work/money). Therefore, a limited military deterrence is necessary for a proper balance between lower upfront military costs and the ability to back a war with enough firepower, and sustained public support, to win it after you are attacked.

That statement represents a *good* interpretation/side of man. The bad side/interpretation would be that mankind is such a lying, killing, power-mad species that we've made war in practically every corner of the globe while plundering the whole Earth, all done with no thought to our fellow men--or even to our own grandchildren. While that is also true, maybe we can change with time. I pray for peace with Osama's followers, really I do. I understand that it will be a generation before they accept peace.

I'm not hopeful this change will come from the top; and that is what this book is really about. The Revelation was given so that you, the reader, will be aware of what will be beneficial for you and your family—without violating God's love or your neighbor. When enough people do that, change at the top becomes inevitable, but not before then.

The Revelation rules of war for the common human are not complicated.

> 1. Fight for your country with bravery when it is attacked, and support those that serve if you cannot. Go into war with the full understanding that it is the dirtiest business your D-brain can do. Don't let your C-brain run your mouth when you return home, talking about things you don't need to say. Never disrespect your war service or those who served with you.

> 2. Support a strong military deterrence so that attacks against your country occur as infrequently as possible. Be mindful of the fact that overspending to eliminate attacks completely is impossible, and therefore only weakens you economically. Reasonable and cost-effective deterrents should be the goal, not the closing of every possible avenue of attack. If we cannot stop North Korea and Iran from getting nukes, and I argue the horse has left the barn in both these countries, why fight containment wars at all? MAD worked with the Russians for 50 years, it can work everywhere.

> 3. Support the war opposition when the government with which you share power uses war for any purpose other than self defense, unless

you have served (then see rule #1).

4. Think of allies willing to die for us, as one of us.

5. Think of all others as the enemy.

This makes you an enlisted man in WWI and WWII. A war opponent during: Korea and Vietnam. An enlistee for Gulf War I, and a supporter of the Afghanistan invasion--but not Iraq. During the Cold War with the old Soviet Union, you would have built 5,000 or 6,000 nukes, but not 30,000. It could be cleaner, but overall, not a bad scorecard. **More importantly**, this is a system that would have put you on the correct side of each of those conflicts right from the start, not after 20 years of debate.

The next big question for world order involves China and the U.S.

The Chinese population is happy and proud of their accomplishments. They are also a proud and strong people historically. They are becoming highly educated and wealthier through the awakening of the population's C-brain forces. This will touch off many power struggles within China as the C-brain forces battle for a piece of the power-sharing pie with the D-brain forces of tyranny. *If the U.S. interferes in China's struggle, even at the edges, it will be drawn in and suffocated.*

The Chinese people themselves must work out the Chinese people's ultimate power-sharing arrangement. This must be done within their own country and within their own system, even if that process necessitates a civil war. The people in China will not trust their new government if their democratic evolution happens any other way.

Currently, the American government does not understand this because the American people do not understand this. We need to understand this in order to resist the temptation to interfere in this process. Any interference on our part will create problems for the entire world.

God is love

Book II, Chapter 10
More on Racism, Convicts and Immigrants

This chapter is the last in Book II for a reason. It is tied to all the preceding chapters and the following Book III, but it is not necessary to have read and understood any of that material in order to understand this one. Furthermore, this is one of the chapters, as mentioned before by me, where I need you to suspend the way you currently think and open your mind to the fact that metaphysical truth and physical truth co-exist, and in fact, are *co-mingled* here in our world and the Non Singular Universe we live in.

To watch Bill O'Reilly and other anti-immigration proponents on television, you would think that the Mexican immigrants who came to the U.S. over the past 20 years are a great danger to America. This mistaken idea is based on faulty C-brain logic and a healthy dose of D-brain fear. People fear these immigrants, but only because they mistakenly believe that they cost the American economy, the American worker, and the American public money. In reality, the immigrants are the most productive workers in the economy: low wage, low public service demand, and hard work in areas of the economy that would cost employers more money to replace. They are by definition high-productivity contributors to the economy relative to both their consumption needs and the cost of replacing them with non-immigrant labor.

Remember the gas cloud theory of all economies. Work the immigrants don't perform either goes undone or is done at higher wage. Both options will cool the cloud, making less energy (wealth) available to all. This is simply the result of the immutable laws of thermodynamics. Any study showing that they cause a drain on the economy is deliberately untruthful or poorly done, usually by people or institutions with an agenda. Check not just the conclusions, but also the methodology of any study. Most construction company foremen across the U.S. will admit that without immigrant labor, mostly illegal, the actual number of construction projects built over the last 20 years would have been severely impacted. Additionally, the cost of each project completed would have been much higher.

Ironically, Republicans and conservatives know this. President George W. Bush knew this and spoke about it many times while Governor of Texas. Governor Bush frequently talked about how important Mexican labor was to the Texas and U.S. economy. He talked about solving the immigration problem by building a middle class (CHUs) in Mexico -- **and** *he was right on point*. But as his standing slipped with the neo-conservatives in the right wing of his party, subsequent to becoming President Bush and after 9-11 and Iraq, he turned his back on this principle and began building the national disaster that immigration "reform" has become. As I said earlier in this book, conservatives are champions of the productivity principle mostly through lip service, not real commitment. They do not understand or embrace the truths behind it.

If a city or town needs a road or a bridge, roads being one of the highest non-energy productivity multipliers of all time, and the construction company can't build it (no labor), or can't build it at a competitive price, then everybody suffers.

For the past 30 years, most proposed public infrastructure projects in the U.S. have suffered from a costly C-brain lie. People convinced themselves that the "No-build" option associated with land use options was the best option. If a "No build" option had been exercised when the U.S.'s current infrastructure was built, there would be no major cities or towns where we live and work, and *are the most productive*. Most no-growth advocates would cheer this situation (e.g. no more road building), until they went hungry. At that point they would break into your house, weapon in hand, looking for food. Don't believe me? That's what happens when economic systems collapse and people are hungry.

Modern politics has produced a bizarre and destructive alliance between the Liberals, who are Collectivist in nature and therefore oppose growth, and the Conservatives, who talk about productivity but frequently vote against it because they don't understand its importance. Furthermore, they both stay in office primarily from rich constituents/special interests who individually profit from monopolies and "special situations" at great cost to the economy as a whole.

People actually believe that if they stop building roads, landfills, and incinerators, waste-water treatment plants, bridges, and additional housing, everything will magically work out. When a facility finally gets built over the lawsuits and bureaucratic impediments, the legal, governmental, and graft "load" on the project sometimes exceeds its construction cost. The "Big Dig" in Boston is just one, classic, modern example of this.

In highly congested areas like New York City, the "no-growth" belief is so popular that *they haven't built a major new bridge or tunnel* onto Manhattan Island in 50 years! This occurred despite the obvious safety and security issues created by a lack of routes out of the city if a crisis were to occur. This belief also causes massive daily productivity losses suffered by the *entire NYC work force* as they crawl in and out of tubes and bridges that are too congested to move traffic and people efficiently. Master-planned roads like the Inter-County Connector in Washington D.C., originally intended to be part of an "Outer" beltway, instead became a lightning rod for discontent with the quality and quantity of growth. The 20-year delay and extended "environmental impact studies" and lawsuits has increased its costs by hundreds of millions of dollars.

Viewed through CLOT, this is the equivalent of taking the hottest part of the economic gas cloud, a productive area that generates the maximum wealth for all of us, and chilling the cloud by pouring cold water on it. New York City has generated enough toll revenue in the last 50 years to rebuild every bridge completely, replace the tunnels with modern ones, and add enough new mass transit to modernize the city into a transportation marvel. Instead, that wealth was siphoned off while the infrastructure itself has been allowed to rot where it stands. This is lunacy. This, metaphorically speaking, is equivalent to siphoning gas out of the car pool vehicle everyday, selling it, and then failing to get everyone to work as a result.

People defecate, urinate, and generate trash -- everyday. They must travel (mentally or physically) to work to be productive. They must live in housing to raise healthy families. All efforts to resist these activities, or to shift the burdens to communities that are poorer and more willing to take the trash, refuse, or radioactive material generated by the politically stronger, richer communities, should be illegal.

Manure used to be a large "environmental problem," but 1,000 cows can produce enough methane to power 300 homes. We could deploy a low-cost, low-tech energy-recapture process that is ready to be used today. We have done this in a few isolated instances, when enterprising individuals have blazed a path -- usually *against the combined forces* of government and the power industry. The current politics of the power industry will kill us through low cost interest programs and energy credits *to promote technologies, which only use fossil fuels.* This is frequently coupled with outright resistance from the utility companies to any non-fossil fuel alternative to producing electricity.

> When we unproductively shift rather than solve our environmental, transportation, and refuse problems, we inevitably suffer the loss of economic growth associated with these transfers. This "cost of shifting problems" increases and escalates the challenges facing us as a species.

We need energy solutions that don't involve sucking the remaining, relatively small, amounts of oil out of the Earth. We need cities with moving walkways through them that can transport people at 18 miles per hour (I have included a plan in Appendix II) so that the vast majority of people moving about each day have to walk, get some exercise, and don't have the energy inefficiency of pushing 4,000 pounds of steel with them to do it. Cars were highly productive, but our limited fuel supplies, obviate the logic of moving that much additional weight in order to move a relatively light human.

When a newly industrialized country, like China, becomes a "first-world modernized country," they need to do it in a way that doesn't pollute their own environment. On its current course, most of China's energy needs for the next 100 years will be met by burning coal. This is potentially the worst solution possible for the human species in general and for the Chinese people in particular from an economic/environmental/productivity point of view. This is only one example of the many decisions mankind has made that are beginning to defy the mathematical odds of species survival. Poisoning our environment prevents future productivity improvements thereby forever linking the two.

Immigration Redux

Solutions to these and other challenges that simultaneously threaten our environment and economies will require humans to embrace unity, porous borders, and productive behaviors. This returns us to the American-Mexican immigration phobia. The Mexicans descend from people that were here long before the white man arrived. They work hard and make less than Americans. In many cases, their immediate families are already here. Furthermore, they came, 20 million strong, practically by invitation from U.S. employers.

What changed so that supposedly intelligent people say: "Now they must go"?

If immigrant expulsion becomes national policy, the sin of anything other than assimilation will have a great price attached to it. If Lincoln were alive, he would be weeping at how little we learned from him. Lincoln would be incredulous that we would spend a fortune attempting to gather up all the hard-working, decent, God-fearing gardeners, chicken-pluckers, construction workers, and cleaning ladies of Mexican descent in order to deport them.

There is much work to be done. We do not have the native citizen labor-pool that has the physical capacity or the volition to do all the labor we require. If we embark as a nation on a necessary capital improvements building binge, can we afford to only employ inefficient union labor? Will we spend the money we desperately need to warm the gas cloud of productivity in a less than optimally productive manner? If the construction projects in that list don't use a significant amount of Mexican labor that will be the case. Any truthful project manager in the U.S. who has built a large capital project in the last 20 years will attest to this. Instead, I fear we'll hear a constant stream of hateful diatribes, which appear C-brain protectionist, but are really D-brain-racist at their core. With great C-brain denial, many seemingly intelligent people justify a deportation plan because Osama bin-Laden, a radical Islamic terrorist, blew up the world trade towers. Truth and logic play no part in these racist choices.

Convicts, and Racism, or The End

After the race riots in the 1960s, President Lyndon Baines Johnson launched his now-infamous war on poverty. Many, many things about this were dangerous and destructive, but none more than the Aid to Families with Dependent Children (AFDC) program, which allowed poor families to get aid based on three core principles: they had to be out of work; they had to have dependent children; the father could not be living with the mother and children.

If you have been paying attention while reading this book, you will realize exactly how poisonous this combination turned out to be. There is nothing the government, white society, or "space aliens" could have done that would have been more destructive to blacks, and poor whites, participating in this program than to combine these three elements. Even outright war on this population group would have been better for them. At least then they would have known it was war and fought for their survival.

The family is the only structure with which the human brain is designed to succeed.

In the 1930s, and 40's, before AFDC, white and black marriage rates were similar. Please look it up. These poor black and white communities were BHU-heavy communities prior to the widespread implementation of AFDC; hence, the aid was directed there. But because of the restrictions racism had created, despite it being over a century between the Civil War and the 1960s, this AFDC poison dart was aimed primarily at the blacks. Prior to AFDC, though heavily discriminated against, blacks, in general, were healthier because their families were intact.

Female brains, by design, will happily accept food/deer meat (AFDC payments) for their children (D-brain programming!). If they have to "move" the males out to get it, they will. The "displaced" males don't actually leave the community; they stay close enough to the females to satisfy the sex urge (D-brain) but are un-evolved by the maturation process, which *only begins for males when they* become *responsible* for children. This made the mothers dependent upon society *and* upon having more children to get more aid (deer meat). Mothers were simultaneously prohibited from working. The males, socially immature and fathering kids with any receptive female, developed none of the responsibility programming that tempers D-brain hunter/killer behaviors. Doesn't this sound like Harlem, Watts, Roxbury, and Southeast D.C. to you?

Blacks did not create the ghettos, drug culture, "gang banging," and the hundreds of thousands of black families with two, three and four children at home by different fathers; the U.S. government did. Politicians who do not understand how the brain works undertook and championed these social programs. They did more damage to blacks than slavery. Unfortunately, ignorance is not bliss--it can mean death. In this case, ignorance was racist and nicely covered by denial and even pushed on by pride--another form of denial.

"The Road to Hell is Paved With Good Intentions" = A Big Lie

This expression is a lie—a big lie. Since God *is* Love, intentions count for a lot. Results follow intention. If your intentions are good, and you are truly acting with love, and motivated by love, you are one with God's plan. The problem is, the AFDC era politicians were not acting with love. They wanted to quell the unrest caused by Martin Luther King's death. They wanted to "buy off" and quiet down this voting block with social programs. They found a formula that would do it regardless of how much harm it caused. Please read just one of Senator Daniel Patrick Moynihan's book's on this subject.

Mr. Moynihan *was* a good man, and a big-time Liberal Democrat. In fact he was one of the architects of the war on poverty in this country. Even though he didn't understand why it was happening (the C- and D-split was not apparent to him), he did understand, later in his career, just how destructive these welfare policies had become to family structure in the black population. By the time of his realization, he was unable to do anything about it.

Both Liberals and Conservatives find it difficult to change failed programs in the U.S. because it requires something modern money-based politics forbids —being perceived that you are attacking your own political/monetary base which will lose you money, influence, and power. The public got tired of the crime and chaos and enacted a special set of laws aimed at punishing this "ghetto" behavior. They proceeded to lock up one million young black men. This did not happen 50 years ago. These men are still in jail in the U.S., where they will have to stay because many of them *are* dangerous. But locked up right along with them are hundreds of thousands of other young black men *who weren't violent,* but were incarcerated anyway because of the inflexibility of the drug sentencing system. President Clinton's failure to correct this problem, despite his personal understanding of this issue and his administration's detailed and widespread knowledge on this topic, perplexes me.

Both groups of inmates, violent and non-violent got there through a process the U.S. government spent billions to enact -- a process that perverted the brain's desire for health and happiness, and used it, to destroy the family. Family is the only institution that can reliably produce productive, non-criminal behavior in super-predator male killers (all of us). Where does this end? The black, or poor white man, leaves jail with a broken brain, a criminal record that makes it hard for him to earn, no family support system, and yet is expected to be productive and keep clean/sober, or else go back to prison for an even longer stretch. The slaves dropped into the ocean off of the slave ships with chains around their ankles received better treatment; at least they were allowed to die with dignity while resisting their captors.

If you are one of these black men and have arrived at the point where you are reading this book, the rest of this chapter is for you. I am talking to you and nobody else.

The Path to Happiness...and an Apology

First, I am very sorry for what my people, white people, have done to you. I know we are all one people. Second, there is hope for you, but it will not come from anywhere outside the mind reading this sentence. You are your only hope. This is no different for anyone else of any color at any location in the world.

Take strength and solace from this, Self-dependence actually gives you complete control over your life. Any other solution to your problem — would not.

Third, the idea that any environmental boundaries or influences can impact your potential happiness is false: false for a moment, false for a day, false for a lifetime. Read this book and read it again until you understand that completely. Your brain is the key to your happiness, your productivity, your closeness to God, your relationship with your mate and children. Those things

I just listed are 100 percent of your happiness. If the things that currently make you happy are not on that list, you just think you're happy.

Fourth, the answer is: God is (equals) Love.

He wants you to motivate your actions by love because He *is Love*. We are playing in His sandbox - the NSU. That means finding a suitable mate and making a family. *This does not require marriage. However, it does require commitment, which is easier to do in marriage.* A committed relationship/marriage with the woman with whom you are fathering children, **and then** taking responsibility for those children, is the only way to produce long-term brain satisfaction. This process encourages you to make love-motivated decisions, happiness and personal success with follow.

If you are not involved in the life of your children, if they don't know you love them by virtue of emotional support, financial support, and daily contact, then you will end up unhappy -- and they will probably end up unhappy and dysfunctional as well.

God's assignment to you is to work as productively as you can every day and make no secret of your desire to do so. *Beware: making money and working productively are not the same thing.* Money earned by productive labor is more wisely consumed than money earned by any other way. So, while productively-obtained money is healthy and constructive, money obtained by any other means, though it appears C-brain equal in all other respects, is much more likely to be consumed unproductively.

Unproductive consumption is emotionally destructive and ultimately corrosive to the productive ability of the human. For example, criminals who prey upon others to make money don't work productively. Hence, their dreams, their families, their businesses, their friends, their freedom, all come to a bad end. Many lawyers also fit this bill, in case you heroin dealers are feeling slighted. I know the following because of the Revelation. *God's promise to you*: If you do the simple and obvious "good for you" things as regards work and family, He will protect you, love you, reward you, and make you happy.

Any other path you choose leads to destruction, hate, violence; and a loss of your mind, dignity, and your children. I do not have to prove this; simply look around with your eyes open. You will see more evidence than I can possibly present here. For those of you just reading this little section—this will sound like one more God lecture. For those of you who have managed to read the book to this point—this will make sense in a way nothing else ever has. Which path you take this moment, despite the road you've traveled to get here, is all that counts. Once you accept the truth that all men are killers and liars, and that humans are instinctively driven to mistrust, hate, and mistreat you, you must also accept the fact that *everybody* faces this challenge. Every race and every person suffers *and* benefits from it.

Furthermore, anybody that overcomes these obstacles does it the same way, because God is Love, and His universe is the one we are living in mathematically, physically and spiritually. We have free will to make choices and choose a path. Put another way: Each of us uses our unique ability to convert volition of will (choice) into an electrical, chemical, and ultimately a material change in our brain. These changes then affect your relationship to everything else in your world. As described in the next physics chapter, it is all connected. It doesn't matter that you don't feel anything when you pray. It doesn't matter that you think believing in God makes you weak. I urge you to look around, but not at the charlatans screaming the name of God to press *their* cause, and for their financial enrichment. What they are doing, by its very nature, is self-aggrandizement and self-enrichment.

> **Look closer, not at the preachers, popes, and politicians masquerading as gate keepers to God's message, but at the one person in your life that was there for you—that cared about you, sacrificed for you, forgave you, made your lunch, and put something special in it for you, that apple or orange you liked but no one else remembered. Look closely, and you will find something special in that person's message to you. Love. I know this exists because God is responsible for these people, just as you one day will act like this to others on his behalf. God is Love. He cares about us and He sends us information and options, choices based on love for every decision we make.**

I don't know which prophets were divine and which ones were not. But, I suspect they all were divinely inspired. Buddha, Christ, Muhammad, Moses, Lincoln, Martin Luther King Jr., Mother Theresa, Gandhi, Einstein, and Pascal: these prophets of love, and others, all had something critical in common. So do you. That's really the key concept for you. You can hear God's voice in many places once you've learned to listen for it. What did all these people come to see, teach, and live by? *God is Love! Only by recognizing this can we choose it and succeed in generating happiness for ourselves, our loved ones, our communities and countries, and even for our species.*

If you have a technical argument against this, please carefully read the Unified Field Theory chapter, which is next in Book III. There are no technical paradoxes in the way. It is possible for God to be infinite, conscious, loving, and in touch with every single one of us, every single atom. We are finite, subject to space and time, mortal—but not without purpose if we choose wisely. He doesn't want any of us -- black, white, yellow or green -- to choose hate. I will admit I don't know the reason for this. The Revelation is clear that He chooses love and *is* love, but nothing in the Revelation told me *"why" He made this choice*. His self-aware creatures, humans, can either choose to motivate their choices based upon His force of love that will help us through this life, or we can choose alternative motivations that all lead to the same wrong choices and poor outcomes.

This free will to choose love is one of the most important consequences of the Unified Field Theory, a mathematical description explaining how all energy forces and matter are related. Prior to the Revelation, I was searching for a system of rules and principles – a balance really -- between what would "get results" and what I felt I was "supposed" to do. I was looking for a way, a path through all the conflicting information I possessed, to consistently make decisions that would produce good results and not harm others, or myself, and not compromise my ethics. I failed miserably every single time. The end was always justifying some means—and both failed to make me happy.

Although I didn't see the complete futility in my actions until the Revelation, I did notice two things. The first was how little good I achieved. The second realization was that the rules, principles, standards, laws, customs, and beliefs, swimming around in my head had become so complex, so circular, and so relative, that I had no *clear* course.

Through the Revelation, I had my first glimpse at a startling truth. It isn't the intellectual depth of our deliberation on a subject or the quality of a decision that actually makes the difference—it's our motivation as to *why* we make the decision. Your decision can result in the same action as another—the exact same action; but, if your action was completely motivated by love, you will get a result that differs totally from someone performing the same action without this motivation. Even scientists applying the scientific method are subject to this rule. The ones free of conflict, motivated by desire to find truth, have a statistical chance at finding it. The ones motivated by a certain outcome, to support a product or technology upon which they are dependent for income, stand little chance of revealing the truth.

"How can this be?" you say -- because God *knows* our motivation(s). Despite being given the correct answer prior to your every decision your denial mechanism is strong enough to lead you away from the truth. He is supplying your brain right now with an alternative from among the choices you are pondering. There is only one choice He is providing…it is the one motivated by love. Therefore, God knows when we do something for love, and He knows when we don't. In order not to motivate our actions by love, we must first reject the option He offers.

To make love-motivated decisions, is to choose God himself. God "lives" in the Singularity and is love. The Revelation revealed that somehow our intentions, emotions, and actions of love accumulate in God/Singularity/Black Hole. This is not spiritual mumbo-jumbo; it is a scientific fact, and not an insignificant one. Ordering the Universe to work this way took imagination, brilliance, care and a desire to know what each atom is up to--as well as to use this process to evolve the Singularity. Since the thoughts and actions done with love are part of His plan, they are fruitful, productive and move us forward—toward Him.

Remember He really isn't a male, He is something else, a *combination* of all of us, a higher consciousness, and more.

The Revelation doesn't predict that all other thought will fail. Free will is dependent upon the existence of alternative motivations, hence its implied sustainability. But, God is love and its ingredients of sacrifice, compassion, patience, productive work, forgiveness, loyalty, and family/common goals. (Incidentally, the way Christ and the other prophets of a love-based philosophy describe these properties spiritually is nearly identical to their mathematical presentation in the Revelation.) Therefore, we only achieve real satisfaction and success in life through the combining of these elements.

You probably have made a number of fair, reasoned, decent decisions un-motivated by love, and watched them fail. You probably have prayed and watched the prayer disappear without answer. These events unfortunately mislead us all. But, I know one thing you have never done. You have never acted completely with love, completely following His voice that is in all of us, and had it fail.

I *thought* I had acted with love as my only motivation hundreds of times. I had acted logically and without remorse. In reality, I was just an atheist, in denial about all around me. Upon seeing what love really is, I have come to see how I had badly handicapped myself each step of the way. You can be a "good" person deep inside and make this mistake daily because of denial. Accepting that God exists is acceptance of a higher power. This process is difficult for the well-intentioned, intelligent, educated C-brain. *Almost* motivating your thoughts by love, is still no good

---

Discovering how to see the path that is motivated by love requires losing our conceits. Conceits are those closely-held obstacles of false information, beliefs, emotions, thoughts, etc., that all of us have placed between God and ourselves. **Love is not an easy choice, because it is a choice we instinctively recognize is *not* ours.**

We want to make our own choices. We want to be with God on our terms. And, He will let you. He will let you make bad decisions as many times as you want. You won't be happy. You may end up in jail, or rich but twisted—and He will let you. Every single alternative to love can be its own "God" in this universe; the choice exists.

What if I am not Christian or Muslim, you may ask, does God love me too? What about the babies born in Africa, dead from starvation before the age of one--how does a loving God allow this? Don't be fooled by the twisted logic of these or other false God paradoxes. The Revelation shows God is a black hole, the Singularity. The Singularity, a black hole, *is a place where time and space have no effect.* That dying child and every child like it who ever lived, as well as every creature that ever experienced love, is part of God himself.

We all die, some more horribly than others, but this transition is part of His plan—a plan that expands to include you when you act with love regardless of your religious beliefs or excludes you when you don't, regardless of your religious beliefs. If you choose to kill another, or hundreds, or even millions, that choice will be open to you. God offers you this in good faith—it cannot be any other way. To prevent horrendous acts from being an option would make free will -- also known here as the second variable -- a sham.

When you block food production and/or aid shipments from being imported, thus killing children by starvation, you are killing human souls with great meaning to God. God himself experiences that evil, something I didn't see before The Unified Field Theory. There are three elements necessary for God to judge our choices. As an aside, this same process is required for evolution to work.
The three elements are:

1. God (SU) must have perfect knowledge of all the material circumstances surrounding a decision.

2. God must have perfect knowledge concerning the motivations of the decision-maker/problem-solver.

3. God must supply the correct decision/solution to the decision-maker prior to the moment of decision. (*This last element must be present because it would otherwise be impossible to negatively judge a decision-maker for not taking a path he/she did not know was available.*)

All of sudden, Christ's advice to tie a millstone to our neck, and throw it in a river before we hurt a child, looks chillingly like the easier path than harming the child. Don't worry about the children you can't feed in Africa; God has them under His wing. Instead, worry about the ones you love and *can help* but aren't currently because you don't motivate your actions toward your own family – by love. To the black man in America; white men like me, put your ancestors in chains and murdered them by the score. They have discriminated against you from day one here. They can, do, and have done, all this to you and your ancestors. But they can't take happiness from you because happiness is a choice you make in your own head when you choose love and family.

What Mankind Needs to Do Now

These final thoughts before Book III are from Karl and Dr. Richards and are for everyone, black and white alike, man and woman alike. We are all human, and we will only survive as a species if we choose:

<u>Love</u> *expressed through mutual cooperation and worldwide economic expansion.*

This includes setting up economic systems that use competition to spur innovation and productive behavior, while not allowing those same competitive strategies to enrich just a few or create monopolies that would undo the good created by competition in the first place. It is **only** in this manner that the high productivity numbers necessary to meet the coming needs of all can be met.

<u>Love</u> *expressed by tempering D-brain greed, racism, and violence.*

You are a killer. Acknowledge it and embrace it, as the Founding Fathers did. They split up government power into three separate branches of government because they knew man was a killer, *not because they thought he would use power wisely.* It's possible to care about someone, love him or her, compete with him or her, work productively with him or her—and not kill him or her. But, we have to establish systems that realistically incorporate the fact that this (killing/war) is the alternative reality if we fail to achieve prosperity with productivity. It is only in accepting, caring about, and serving others that we can enhance our own quality of life.

<u>Love</u> *as expressed through nurturing/protection of resources and renewable energy.*

For the foreseeable future, we are stuck on this rotating planet. We must treat its resources as if our lives depend upon it. I know this sounds ridiculously obvious; but surprisingly, our combined behaviors led by mass denial are shockingly shortsighted. We are fragile creatures with very specific tolerances to temperature, humidity, chemicals, and radiation. Additionally, we are way too big. So far, only creatures smaller than a vole (a little mouse) made it through the majority of our planets past extinction events. That means we as a species must actively anticipate *and prevent* the next extinction event. The only other survival option is to quickly evolve into *smaller* creatures. Extinction is much closer than you perceive it to be.

Most importantly, if we wish to thrive as a species:

It should be our goal to choose love as the motivation behind every C- & D-brain decision we make, action we take, word we speak, and every thought we dream about. It is the choice that comes from God, not from within us, and that is one way you will know it -- it is offered.

BOOK III

God is Love

Book III, Chapter 1

There is a God

The Metaphysical Explanation of the Unified Field Theory

Book III is last for a reason. Up until now, we've dealt with how important our evolutionary past is to our current behavior and thought patterns. We then explained how those behavior and thought patterns shaped our societal, international, and economic relations. We have debunked more than a few false beliefs. We've swept aside false ideas that religions use to control people and stay in power. We've explained how corporations, unions, and the government are manipulating the economy to help themselves—while taxing, starving, and ultimately killing you.

Does all of this revealing thought lead us to conclude that we are simply riding along on a planet ruled by super Darwinism and killer humans? Is our world created by a God who is great at math, but did not leave us with any clues that describe who we are, and how we got here? What are we supposed to conclude? Let's approach this emotionally-charged topic one step at a time. Before you atheists get upset, remember your promise in the first chapters to stay open minded.

There is a God. Through the Revelation, I can prove it. What's more, I can show you how to find God, right now--in your life. God *is* Love. I know this sounds silly, because what is love? Love is the will to make the right choice from imperfect information. To love is to work productively, forgive harm, sacrifice wealth for future productive gain, risk our own death for our family/clan when they are threatened, deal with our fellow humans in good faith, communicate truthfully and be aware of our temporary place in the environment. And, most importantly, acknowledge a higher consciousness.

If our information is imperfect, how do we know which is the love-motivated choice? We know this most important fundamental truth because: we all hear the Voice. The Voice coming from another point of view, but still inside our head, guiding us, *Come on, admit it!* You are not crazy to admit that you have another side; a side guided by thinking that is not rational and yet is not *irrational*. We all hear the Voice that offers us a love-motivated choice whenever we weigh a decision. This is the voice of love, the voice of God. In the beginning of this book you were promised an explanation of the Unified Field Theory. It will be explained very shortly. *Since God was integral to the beginning of the universe, The Unified Field Theory is really God's creation story told through mathematics. These mathematical rules are the principles He used to create our universe, the Non-Singularity Universe, or NSU.* Therefore, in explaining the Unified Field Theory and the universe in which we currently reside, it is necessary to reveal that God is in this formula. Dr. Stephen Hawking asks the essential question: how does the fire of an idea become actual nuclear fire? We know this can be done; the fire of an idea literally becomes matter in the brain. This then *must be* possible on every level. An idea divided from an infinite energy source would not be a different amount of matter altogether but the same exact idea, energy or matter.

The rules of mathematics/matter/physics/energy/evolution/CLOT cannot be contradicted. Why? For one reason: they are God's rules for the universe. People often hope for a miracle that will defy physics (the rules). Don't hold your breath! You will *never* see anything defy these rules—because of their origin. Einstein was right. He made calculations (perhaps with the help of a revelation?) about the Universe's Standard Model and the balance of mass/energy through the use of what he called the "Cosmic Constant." Why is a cosmic constant necessary? Importantly, Einstein made us think of time differently than we ever had before. According to him, time wasn't just an abstract idea such as beauty. Rather *time was an actual element in the construction of our universe, a real mathematically-related force.* He was proven right.

This should make it easier for the reader to understand that there exists another force in the universe (other than time) that is not just the abstract idea it appears to be. That force is love.

While it is difficult to think of time in a manner similar to the 3 spatial dimensions, it is even harder to think of love as more than an abstract idea. But love, like time, is a real force--not just an emotion or an idea. The triangle included at the end of this chapter, the formulas, and descriptions, about what happened in the moment before the Big Bang, the combined law of thermodynamics, Pi, Prime numbers, and the *Peace Machine:* these all lie perfectly on top of each other in the Revelation.

According to the Revelation, Pi and Prime numbers are the keys to the mathematical description of the universe, then, now, and in the future. They are essential to understanding nearly-perfectly-compressed material that is, by definition, spherical in shape. Everything beginning in the Singularity is a definition, except Uncertainty. Everything in the NSU is a derivative, except quanta, which is defined as a photon and the only thing unaffected in the NSU by Uncertainty. Our universe's rules are dependent upon the rules defined by the "SU." Therefore, mathematic and physical laws define everything subsequent to the "SU." In other words the mathematic and physical laws (including Uncertainty) selected by the SU completely and thoroughly describe our universe, the NSU. Hence, this makes a Unified Field Theory possible and guarantees its existence. Without the SU consciously defining our laws, the laws would be random and no Unified Field Theory *could* exist. This is a different principle than the idea that there is a design, therefore a designer. Although that happens to be true, *it doesn't have to be mathematically*. This is a recognition that since all math is related and the fields (energy, matter, weak and strong force, etc.) in our coordinate system and the math are inseparable, the presentation of the different fields like matter and energy that we can observe--was by conscious choice. This does imply a designer.

Physicists are attempting to reconcile the aspects of Relativity (Certainty) and Quantum Mechanics (Uncertainty). These principles run through everything. They are looking for a small Unified Field Theory. In reality, there is only a big one. The next 3 pages will describe the metaphysical side of the Unified Field Theory. The following chapter describes the mathematics. They are one in the same, just expressed in different languages. I will start with the original love triangle:

    Father
   Mother-kids
 Grandchildren-family

The triangle above is the first triangle, one that nests at the bottom of all the others. It is the foundation and the one to which you need to pay the most attention. **It is the supporting base.** If the head of this triangle motivates his/her actions by love, these actions will never be inconsistent with the D-brain goals, healthy C-brain goals, and the unifying force in the universe— love. Each subsequent and larger triangle is then constructed on top of this one, and as part of it must *share* all its fundamental characteristics. A model of this larger universal pyramid is on the following page. Note that in the bigger triangle the President of the U.S. (or the King of Saudi Arabia!) acts as a Father/leader, filling this role for the country. Love-based decisions necessitate turning our back on all other thoughts and actions not motivated by love. Since we have free will, we will always have a choice in every decision, every action, and every movement. We are constantly being offered the love-motivated choices concerning our decisions, emotions, actions, and thinking. We can feel it, maybe only just slightly, but we *can* feel that love choice. Sometimes it is found in that second of remorse we feel a moment before we do something we shouldn't. Once we understand God is love, and is at the head of the final and top triangle, the Revelation provides a roadmap about what to do in order to unify our actions with His force. Everyone has an internal indicator within, a needle in the moral compass, even the mentally ill. That is God's guidance, His thoughts, and His offering to us, of a love-motivated and love-based choice. What, then, is free will?

> Many currently believe free will is the ability to make a choice from many options. However, according to the Revelation, all choice is without consequence except one: the choice to let God change us—make us more like Him. We can join the love force moving *in* space and time on our side of the universe, the NSU. We can do this by choosing what we know is the love-motivated choice.

This is the essence of free will. Do we choose to motivate our actions and thoughts by love—which puts us in God's plan? Or, do we practice any other alternative: which by definition is in opposition to God? This first triangle nests inside others; some of which have already been completed by man, some of which, like the unification of mankind, are not. The following triangle represents God's outline, as expressed through the Revelation, concerning love. When you choose to be motivated by love, your actions become part of His plan. He divided Himself, leaving Himself still whole and infinite.

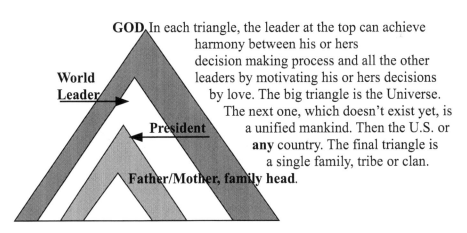

In each triangle, the leader at the top can achieve harmony between his or hers decision making process and all the other leaders by motivating his or hers decisions by love. The big triangle is the Universe. The next one, which doesn't exist yet, is a unified mankind. Then the U.S. or **any** country. The final triangle is a single family, tribe or clan.

This system eliminates relativism and existentialism.

If a father or mother motivated by love works hard to grow food to feed his family, the other family members will benefit, as will the leader of the clan, or country, where he/she lives, as would the leader of the world, if there was one, as will God. In all ways and in every decision, this dynamic is always true. Love-motivated decisions for members of the triangle above us and below us never present a conflict with any other love-motivated decision, or love-motivated decision maker, anywhere in the universe.

In the next chapter, using the mathematical formulas, I will show you how God/SU physically created the part of the Universe in which we reside, which makes this love triangle work. Most physicists are only looking to unify Relativity and Quantum Mechanics. This is actually very odd, since a true Unified Field Theory would have to reconcile much more than this—and indeed, it does. If you've completely dismissed this book as religious psychobabble, next time you hear the Voice, pick up the book and give it another read. This very profound formula (God = Love) has its own space in the Revelation, a spot on top of all the other ideas. It literally appears throughout the Revelation, woven in among the other formulas and coming out on "top" of all the information that I detailed in this book. This interwoven nature is more than just interesting. God is not only a force in the universe we live in; He is something we know about and use in our everyday lives. Furthermore, He/It knows about us, each and every one of us. He is concerned with our welfare. **In the Revelation *He/God/It/One/SU* does not have a sex; I use the male pronoun because I have no better way to express this in the English language.**

Actually, as already mentioned far earlier in the book, there are no *words* in the Revelation at all. The equations and images in it represent ideas, many of which can easily be translated into words while others have taken years of intellectual struggle to unravel. God, Love, and Light, all appear as one symbol/idea in the Revelation. Love isn't a tool of God, or a thought He has. He actually *is* Love. He/love is: a force, an idea, a system of change (evolution), an emotion, and energy (light again) of incredible power. I understand from the Revelation that men and women represent two different parts of the love "whole." In that respect, the religious statement that God made us in His own image is correct…and finally understandable. Every creature that can love is created in this image.

The Revelation was presented throughout this book in the sequence it was shown to me in order to expose the interrelationships of the concepts as they were presented. Unfortunately, there are no words with which to completely describe the linkages. I hope an "intuitive" or "inspired" understanding will occur in your mind, if not initially, then maybe after a few readings of this book. Hopefully, you will further grasp the relationships after you experience or observe powerful life events.

Book III, Chapter 2

The Unified Field Theory

This theory contains mathematical formulas and the resultant truths that reach far beyond mathematics. This may make some people uncomfortable, because they don't instinctively think of math as other than an abstract expression of an observed event or a conclusion. If mathematics defines the rules of the universe, as many scientists and the Revelation agree that they do, then *mathematical rules are absolute. There are no exceptions. The logic in the rules is as important as the result. Matter and energy obey the rules as if the mathematical "abstraction" defining them had come to life and occupied the subjects.*

Having said the above, I admit that this is not how humans developed our traditional understanding of mathematics. We initially grasped math by observing and then writing down rules that seemed to describe our observations. This served us well enough until the theories of Relativity and Quantum Mechanics proposed rules about things we could not directly observe. Oddly, we found the formulas we developed describing the directly unobservable very useful and enlightening. We have subsequently proceeded to prove many of them by observation of the interactions of large-scale (astronomical) events and very tiny scale (quantum) particles. Amazingly, the mathematics of Relativity and Quantum Mechanics held up!

Certain very smart people, like Einstein, were never comfortable with the Heisenberg/Bohr Uncertainty Principle's effect on the equations. In presenting the Unified Field Theory, I will be asking you to do something one of the great minds of all time had trouble doing. Einstein's problem: he was human. We are all human, and this makes math an abstraction for us. Therefore, many of us never really trust it. This mistrust has been helped along by the continual refinement, disagreement, and even complete retraction of certain things we were taught and intrinsically thought were true before "new" discoveries came along.

To mistrust the absoluteness of math is one of our inherited curses—a legacy of our beginnings as creatures so ignorant, practically every abstract thought we had turned out to be wrong. We "remember" this biologically. Without our inherent mistrust of the abstract, we would not have survived the evolutionary times during which our main struggle was with "lions and tigers," not quantum physics and the origins of the universe. The human C- and D-brain is an evolutionarily-derived arrangement that got us through primitive/subsistence times. It is now struggling to help us understand the universe.

If you were an intelligent man unfamiliar with mechanized flight and you found an airplane in a field, you might make hundreds of incorrect assumptions while dissecting it. You would ponder what made it fly, but only discover Bernoulli's principle (the airflow principle that allows lift under a wing to occur) very late in the game. Nothing you could physically observe about a plane sitting on the ground would lead you to this principle directly. Parts, like the engines, might even lead you away from the core principles of flight for a very long time. In the end, the Bernoulli principle would be discovered. The controlled flight of a heavier-than-air structure would not happen without conforming to it. To truly believe in mathematics, takes faith. *Our most respected scientists, the pillars of logic and education, demonstrate this faith every day:*

> -- Faith that an abstract formula can tell you how matter and energy behave.
> -- Faith that to calculate, not measure, the circumference of a circle anywhere in the universe requires an infinite number, Pi.

I need you to take that same leap of faith in order to understand the Unified Field Theory. I need you to suspend your disbelief long enough to accept that the number "One" can't be divided by any number other than itself. This is true here and *everywhere else.*

You can't think of the number "One" like one pineapple or one shipment of grain. I need you to think of a "One" as a special number to which special rules apply. These rules do not apply to a pineapple or a shipment of grain because those things are not really a number "One" at all. The pineapple and shipment are objects containing many other things--we just label them as one pineapple or one shipment for convenience, not truth.

Math and Physics – A Brief History

First a little background, as poorly explained by me.

Our explanation begins with Sir Isaac Newton, a very smart man who developed Newtonian physics and was then able to calculate the mass of the Earth to a result still considered accurate today. His mathematics gave us the ability to calculate how things worked here on Earth to such accuracy that virtually all of the math we need to know about how to get things done on Earth can be derived using "classic" Newtonian mechanics and its predecessors (geometry etc.). In Newtonian mechanics, as in all mathematics, certain strange things happen with Pi and Prime numbers. Pi has great power, letting us predict the circumference, the surface area, and volume of circles and spheres. Pi is a mystery in itself. It has been predicted to be an infinite and random number, and then found to be so through many methods that arrive at Pi through completely different logical processes or "proofs" as they are labeled in math circles. All manner of deriving Pi achieve the same endlessly long but incomplete number.

Pi is a very powerful entity, one that we must look at from many different angles before understanding this explanation of the universe. Pi itself is proof something can be infinite. This is very significant, because if Pi can be infinite, something else can also be infinite. This one constant/non-constant (Pi is both) proves the existence mathematically of both Certainty and Uncertainty. Pi also proves that things, which are infinite, can exist side by side in a universe with things that are not. It is necessary in every math formula known to predict any part of the description of a circle or sphere. Pi is one of the doorways between the SU and NSU.

The above description of Pi sets the stage for a duality in the universe that is central to understanding the Unified Field Theory. In other words, although *we* only observe one universe (our NSU), math tells us there is a second one —*depending upon your place of observation.* This also is consistent with Quantum Mechanics and Relativity, but is not obvious since Heisenberg, Bohr, and Einstein (in his criticisms) did not concentrate on the importance of this ramification of the Uncertainty Principle. Metaphorically speaking, they were fascinated by the plane's engine and missed the importance of the curve of the wing. Pi isn't the only proof of this. Einstein's Relativity calculations concerning the collapse of a star's center (and gravity) to infinite mass and density at the end of a large star's life is yet another piece of mathematical proof that infinity is not only possible—but is happening around us with matter and energy that *was once finite* because it existed here, in the NSU, a finite universe where space, time, and therefore an end to those concepts, also exists.

Pi, by its nature as an unending number, also proves that the edge of a sphere (or circle) can never be calculated exactly — proof once again, of The Uncertainty Principle. However, this uncertainty was not expressed as a mathematical principle (Quantum Mechanics) until hundreds of years after PI was approximated and thought to be infinite. In other words we grasped the infinity of Pi, hundreds of years before the Uncertainty Principle became obvious, although Pi, necessitates Uncertainty in our universe, the NSU.

> Pi does all of the above simultaneously because it *acts* as a constant. In other words you insert it into a formula seeking a description concerning either a circle or a sphere--and presto! You get an *almost* accurate answer.

Our adventure inside abstract math began in earnest when man looked toward the heavens and started to calculate solar distances. When exploring astronomical bodies, Newtonian mechanics were inadequate/inaccurate over those distances. To measure the distances between stars, or to account for the amount that gravity bends light when it passes by a massive object, you need Relativity. Einstein created two types of math to handle the two circumstances he theorized were possible, a coordinate set or CS to handle the universe without inertia, or one free of all outside influence. He called this Special Relativity. He also created General Relativity to handle a CS without exclusion for all other outside influences. These two math's essentially recognize the importance of seeing the universe(s) from two different points of observation in order to make sense of it.

The math Einstein created was fascinating stuff. It predicted strange things such as the ability to convert mass into energy (e=mc squared), the fact that light cannot travel faster than 186,000 miles per second, and that as you speed up relative to someone else—time slows down. All of these conclusions are essential to the Unified Field Theory presented later in this chapter. Much of this was controversial when it was first announced, but almost all of it has now been proven. Mr. Einstein worked out a formula, which he believed described the entire universe—but he couldn't make the equation balance without the addition of a new constant. Einstein called this constant the Cosmic Constant. He later retracted his formula describing the universe, calling the Cosmic Constant his biggest blunder. The truth of that statement is still in question. Ironically, many physicists today believe Einstein's Cosmic Constant was a brilliant piece of insight. Many feel that his Cosmic Constant was a misnomer for the cosmic vacuum—a bit of modern theory necessary for explaining celestial observations that are much more accurate than during Einstein's time. The currently-accepted math associated with the speed of the expansion of our universe requires such a vacuum…they think. Einstein was a very loving man, and his friends were mostly fellow physicists with whom he argued, until his death, over a wrinkle they found in Relativity — its certainty. But, they were cordial and enjoyable disagreements, perhaps a more important lesson than his math. In reality, what they were discussing was a lot more than just a wrinkle. As useful as Relativity is, it didn't work in the micro world (that really, really, really tiny place where electrons whirl in elliptical patterns, shells, orbits, or clouds) anymore than Newtonian theory worked for celestial calculations.

That "wrinkle" existed for reasons which Mr. Bohr and Mr. Heisenberg explained eloquently in their papers. At the heart of Einstein's relativity quandary: *it is not possible to calculate where an electron is located in space, only where it might be statistically speaking.* Furthermore, we would never be able to measure where it is -- since any such measurement would itself change the answer. In fact, Bohr and Heisenberg predicted that any attempt to observe that electron would change its location, behavior, and properties.

This principle is called the Copenhagen Interpretation of Quantum Mechanics, commonly known as the Uncertainty Principle. Its importance to our universe, mathematics, even Relativity itself, cannot be overstated. This principle is at the core of the way the universe operates and is simultaneously at the core of man's relationship with God as well as with himself and others.

Looking back through the eyeglasses of the Revelation, it now seems obvious that whatever process lies at the heart of the universe's creation would also lie at the heart of everything in the universe, including us and our math. Since God created our universe from His/Himself, we are actually made up of His material, not so much a copy as a diluted (literally less dense) form of Him. Things like space, time, matter, energy, charge, the weak nuclear forces, and the strong nuclear forces result from His purposeful expansion of a piece of His universe, the Singularity or SU, to create ours (NSU). Think of time, space, and energy as "fields" separated out from a material of such a concentrated density, that the individual pieces we perceive, as different things, are all integrated into one thing at the place of origin. The theory presented here extends this definition beyond those classic fields like time, space, matter, and energy into every *characteristic* of our universe.

Back to Brain Structure: Function and the Uncertainty Principle

This idea, that the fields are just different presentations of the same thing, extends to our brain, a quantum instrument. Our brain's physical structure is tied into the Uncertainty Principle by design. Immediately after the human mind initiates an action or thought, we make a choice. It is at that moment we exercise free will--that expression of volition that translates, changes, or cancels our response to our initial thought. We can choose to be motivated by love or by any other emotions/needs.

This important paragraph from Book I chapter 2 is repeated here:

"One of the central themes of this book is the concept that a decision is made in the base area of the brain and then makes its way to the "outer" brain, which then approves, rejects or alters that first decision. The Revelation theory postulates that this interplay is constant, continuous, and represents a brain process designed for and capable of allowing the brain multiple inputs via a second variable (a second opinion separate from the first impulse to act which was supplied by the ancient predator/killer D-brain). This second brain (C-brain) is born blank, unlike the predator/killer brain, so it can *learn* information (like reading and writing) which is impossible to transmit through the genetic process."

The exercise of free will is a physical representation of this Uncertainty concept. Math itself, then, by design, makes free will in humans possible. Brain "plasticity" describes how the brain redesigns itself physically on a constant basis as a result of experience. This is the end result of our exercise of free will. This translation of experience molded by conscious choices into the wiring of a flesh and blood brain is a physical manifestation of the interaction between God and man. Our brain matter decides -- out of the uncertainty -- which action to take, what thought to have, and what word to speak. Most importantly, we make patterns of decisions based upon our past decisions, actions, and motivations. *The process observed by Libet (Book 1, chapter 2) is literally a translation, at the quantum layer, of thought into reality.* This process, mirrors on a minute scale, the moment when the consciousness of the Singularity decided to expand a small section of its concentrated entity in order to create our universe, the Non-Singularity Universe, or NSU.

All of this was possible because of the Uncertainty Principle. **Only things that can change can have consciousness**. So, Uncertainty is necessary for change, for evolution, and the consciousness of a two variable entity like humans. Pi then, is not a limitation on the Singularity; its uncertain nature is a necessity. It is a marvelously designed tool that makes it possible to attempt a flawed description of a nearly-perfectly compressed sphere, similar to a camera lens that can never get a scene containing near and far objects in focus at one time. As the lens twists, different parts come into focus and the picture continually changes. This is necessary because the ability to change is a necessary ingredient of consciousness, even in a black hole. It is the Singularity's ability to change that mandates Pi to be infinite and random, precisely because its complete description can never be accomplished.

The quantum relationship is the method by which the Singularity remains in constant contact with every atom in our universe (the NSU) – occurring through quantum mirroring in the atom. The electron oscillates at the speed of light, putting it right up against the NSU's boundary with the Singularity. The electron's function is similar to a human nerve ending, transmitting the necessary data back to the quantum/conscious/entity through a process that is less than perfect in biology, although still impressive. The atom's process transmits *everything* to the SU. How does it do this?

The electron is always at 100% potential because of its oscillation rate. Anything that enters its environment will change it -- Uncertainty Principle again. It is also connected to the strong nuclear force in the center of the atom, as well as through the electrostatic forces between the Proton and the Electron. Changes affecting any matter (or energy, or any field) will alter the combination of forces affecting the nucleus and the electrons thus making the electron/proton combo an exquisite sensor of its environment. The strong nuclear force, in turn, is tied to the quantum mirroring that occurs between the matter in an atom and its anti-matter "twin" in the Singularity. By this mechanism, whatever happens in the electron's sphere of influence is quantum mirrored in the Singularity. *The same mechanism is/was at work:* in the consciousness of the Singularity (or God), at the moment of the Big Bang "division" when a part of the Singularity became the origin of the NSU, at the quantum process in the heart of every atom, and in the quantum consciousness of a human brain.

Imagine a consciousness inside a super-concentrated material, spherical in shape that can change. Imagine it changes a small amount of energy/matter into *less* concentrated material at the initial velocity of 186,000 miles per second *into* a place where time, space, and an alternate state of charge can all exist. It would still be connected to that newly defined, non-SU material, just as the consciousness in your mind is connected to the physical reality of your brain. By extension, the entire NSU is still connected to the Singularity. How can some of the same material exist in the SU and the NSU? Though far beyond the scope of this book's explanations, the answer involves what we in the NSU would perceive as a "time shift." Time does not exist in the SU and was created as a way to construct the NSU by the SU. Our fields are spread out of the SU's material by separating those characteristics from the SU, *only* by time. This is possible because the SU is infinite, and since time does not apply there, what appears to be a great spread of time to us, like billions of years, is an immeasurably small difference between the two universes. Quantum Mechanics thinking by Heisenberg, Einstein, and Bohr about Relativity and the Uncertainty principle reached the point where the public and non-mathematicians basically stopped trying to comprehend this part of physics. This is unfortunate; they quit right before the finish line. It was also the point at which scientists and mathematicians entered their own tiny world of higher-level math, which requires years of study to understand. The idea of Uncertainty bothered Einstein, and quite frankly, he never accepted it. The search to reconcile these disciplines into "One" math, that would describe everything (The Unified Field Theory) began at that time and has never ended. But, a bigger point has emerged over time with certainty (how ironic): Quantum Mechanics is not wrong, and the math generated by it has contributed as much to today's science as Relativity and Newtonian mechanics did previously.

String Theory, Stephen Hawking, and the Black Hole Error

There is a fourth discipline, string theory, on which I'm not going to write much. It takes place at an almost impossibly small place, a place so small that I will use the following analogy to give you perspective. Imagine that an atom's nucleus is magnified to the size of the Earth. A period at the end of a sentence in this book on that imaginary Earth describes the dimensions of space in which string theory applies. We now have four Universal size scales and four types of math: Newtonian, Relativity, Quantum and String -- none of them completely compatible. But, they all have threads of compatibility and continuity running through them, like Pi (good old Pi) and the importance of prime numbers. It should be noted, though, again, that no one--up until this Revelation--has any idea why these threads are there or why they might be important.

Remember our deal concerning faith in math and the number One—we're going to need that soon. But, before I start pulling it all together, I want to mention a great scientist who is alive today, Dr. Stephen Hawking. I mention Dr. Hawking because he has spent a good portion of his incredibly productive life working on black holes. Dr. Hawking described a lot of what is contained in the Revelation's Unified Field Theory, but in bits and pieces.
For example, Dr. Hawking was the first physicist to disagree with the idea that *nothing* can get out of or be emitted from a black hole. In fact, he correctly concluded that the radiation emission from a black hole could create matter that literally appears in space, just beyond the event horizon. Why he did not take this concept to its logical conclusion about the galaxy building nature of black holes—I do not know. On the flip side, Dr. Hawking and Roger Penrose (another great mind) believe that things that exist in our space-time continuum (like a space man or hydrogen gas) can "fall" back into a black hole. For the reasons stated below, this is *not true*.
The definition of entropy derived from the Revelation is different from that described in classic physics. Entropy is incorrectly viewed today as a tendency toward "disorder" rather than a description of the results generated

by observing the physical manifestation of the combined laws of thermodynamics, (CLOT) -- which is exactly what I believe entropy to be. No disrespect to Hawking or Penrose; both are much smarter than I. Frankly I did not even finish one complete year of college, and had to cheat to get out of High School. If this shakes your faith in this book, I'm sorry, but I've become a full-disclosure kind of guy.

Max Planck was one of the greatest thinkers of all time. His thoughts on the combined laws of thermodynamics are quoted below because we need to refresh ourselves with this very important principle before jumping into the Revelation's Unified Field Theory.

> "The second law of thermodynamics states there exists in nature for each system of bodies a quantity, which by all changes of the system either remains constant (in reversible processes) or increases in value (in irreversible processes). This quantity is called, following Clausius, the entropy of the system...
> Since there exists in nature no process entirely free from friction or heat conduction, all processes, which actually take place in nature, if the second law be correct, are in reality irreversible. Reversible processes form only an ideal limiting case. They are, however, of considerable importance for theoretical demonstration and for application to states of equilibrium."

As you will see, this is a pivotal point. The combined laws of thermodynamics apply to our universe, are irreversible, and actually began at the moment space/time was created. The NSU formation out of the SU was/is literally the force behind entropy. However, these laws likely do not apply in the Singularity, or are reversible there if they do exist, as the Singularity is by definition infinite. All this is true while we co-exist outside the Singularity, on the other side of the interface between our finite universe and its infinite one. Our universe *is* subject to the effects of space and time. This makes entropy inevitable for us, **and** all time irreversible by us, *until* the Singularity re-absorbs us. Entropy understood in this way also gives us insight into the "natural decay" of our universe's matter and energy. This observed atomic decay is really just a re-absorption into the SU.

THE UNIFIED FIELD THEORY – EXPLAINED

Enough background, here we go. For the physics aficionados among us–here is the ride of your life!

In the beginning, there was a Universe, but not the physical one in which we currently reside. As a result of our human-centric beliefs and resultant denial, *we firmly (but falsely) believe we are in the universe as it began.* This is an important reason we have never before found the truth about the origin of the universe. If you begin by believing we were in the universe as it began, you *cannot* find the correct answer. We are on the finite side, (or the 2nd side), so the actual concept of infinity or "No-time" is impossible for us to fully grasp, since we always must *consider* the second variable and also literally must live subject to space and time.

The journey begins very simply with the concept of the entirety of all matter/energy (which we call the Singularity) that is infinite and forever present, in existence prior to our universe being created. For this to be understood, we have to more fully expand our definition of the number "One." We have to create a one that is shown like this One (1), or *the* One, but also a *number* 1. One (1) is a number which not only existed and still exists in the Singularity, but that truly only exists there. Think of a universe where there is only one number, the number One, and a second universe, started by the first, continuing with the rest of the numbers, beginning with a 2, and proceeding through the rest of the primes. Our human version of the One (1) is a placeholder! The Singularity's number One (1) existed before the creation of space and time and was the complete Universe by mathematical definition and physical manifestation since mathematics and the physical universe cannot disagree. By definition, this has to be so.

Something cannot be everything, complete, and infinite, *without* being a Singularity. The mathematics in Relativity demonstrates this about black holes. Again, math demands it; and the physical universe demands it. The One (1) is not subject to space or time, nor can it be, because it is infinite. Again the mathematics in relativity concerning black holes demonstrates this.

Current math (physics) describes an *event horizon* surrounding the black hole as a place where time stands still. This is true. It then mistakenly *predicts matter can cross this horizon and be sucked into the black hole.* **This is untrue.** Nothing with mass from this universe can cross the event horizon and enter the black hole. To do this, the matter would have to travel back in time. We already know this is impossible because of the combined laws of thermodynamics or CLOT, discussed earlier. Basically, once heat is lost to friction, to entropy as we defined it, it cannot be re-gathered. So nothing from the universe where that heat is lost can travel back to the point in the time/space continuum where the heat was concentrated before being lost to friction.

Relativity also forbids matter from making this transition, since it would have to exceed the speed of light to pass the event horizon into the black hole. The reason for this obstacle is that matter (traveling slower than the speed of light) could not do that without gaining infinite mass--something it would have to do to re-join the Singularity.

This "Revelation"- inspired deduction about how matter passing the event horizon to join an infinite mass object would behave is consistent with Relativity. It is also consistent with the Revelation's requirement for objects with mass that exist in our Non-Singular Universe to become infinitely massive the moment they attain the speed of light *and* go back in time to where the Singularity (infinite mass) awaits. Astronomers claim to have photographic evidence of matter being sucked into black holes. This will eventually be shown to be erroneous. The One, (1) *is* The Black Hole and the number 1-- to the complete meaning of that definition. It is the actual number One (1), because it is/was the first "proton," although an infinitely large one, and the only one before time.

All Black Holes Are One

If you assume we live in the only universe – this concept breaks down. But, if you can mentally picture a Singularity Universe (SU) creating an Non-Singularity Universe (NSU), as a place composed of material that cannot observe the SU because of its order of creation, then we can understand what we call black holes -- as just doors. The SU on the other side of these doors is the same one! It is like a building with many different entrances when viewed from outside but only one continuous room inside, the original number One (1).

One (1) is not a prime number as we currently define them, because primes can be divided by themselves *and* the number one (2 divisors). Math textbooks admit this but gloss over this problem, leaving the definition of primes in place, even though the number one admittedly contradicts the definition.

The Singularity, the One (1), can only be divided by itself; there is no *second* divisor. The absence of this second divisor is critical to the properties of One (1) as a super prime. Grouping the number 1, in with the other primes was a bad mistake mathematically and logically.

If the One (1) is *only* divisible by itself, this implies/requires that it must have chosen to create a finite sub-set (our NSU) because nothing else can divide it mathematically—or by extension — physically.

If you don't have faith in math here—we lose you.

For our NS universe to be divided out of the One (1), to be changed from a state of infinite unity, compression, energy, and mass – there must be consciousness. If you were guessing that this mathematical principle implies consciousness in the Singularity, you are correct.

Limits

We understand the paradox of living in an apparently finite universe (the NSU) with no provable or even plausible explanation for a beginning. We also understand the falsehood of how our universe could be infinite without the benefit of enough mass or a re-start mechanism. Furthermore, if the NSU was infinite, then things within it could be infinitely small, large, fast, or slow. But quanta exist; there are physical limits on the building blocks of our universe. In fact, they only come in one size, making the paradox apparently deeper, particularly if we cling to the belief that we are in the only universe.

Rarely appreciated for this important quality, the energy level of one photon demonstrates the consciousness of the Singularity. Humans have been slow to recognize this, although I believe the great mathematician Max Planck sensed it. The discrete nature of these energy building blocks *that come in only one size and are not divisible to a smaller unit nor come in any other size at all,* define **quanta**. This fact (like Pi) *also* implies consciousness. In other words, the Singularity has consciously chosen that level of energy, *and only that level of energy*, to move in discrete quanta in and out of the Singularity. A non-conscious entity *could not* make such an arbitrary floor. Matter and energy present anywhere in an infinite universe would be theoretically capable of being further divided into smaller, or combined into bigger, parts. A floor on this division, a place where nothing can be made smaller demonstrates consciousness. It also leads us to the conclusion that the NSU is but a finite subset of the infinite SU. I believe the size of this unit, the photon, will eventually be shown to be the product/result of the spin rate of the Singularity as viewed from our NSU perspective.

There is another obvious limit in our NSU...the speed of light. As we will see later in this chapter, it is another one of the Singularity's consciously defined limits on our universe. Some scientists will term these two decisions as only a mathematical *possibility of consciousness* in the Singularity--which from a purely mathematical point of view I cannot dispute. However, I know the Singularity's consciousness is more than a possibility. The Revelation I experienced, containing this information about The Unified Field Theory, also contained the singular Revelation statement/formula: **God = Love**. Coincidence? You may think so—but I do not.

The mathematically different definition of super prime "One" creates a new class of numbers, of which there is only one: the "One (1)." The One (1) has one divisor, itself. The One (1) must have also been infinite since it was the One (1) and there was nothing else. Therefore, it was everything. It is also a *black hole* since they are infinite. All manner of infinity must be mathematically and physically amorphous, or put another way, "completely the same, or all out of the same." By having charge, the Singularity can be a switch without being a 0 and 1. It uses what we call negative and positive charge, which is just a different state of being of the same energy/material. By definition, the principles and laws of mathematics must be the same, and their effects must be uniform *on this side* of the observation point – the human viewpoint in the NSU. Mathematics were the rules of our universe from the point at which time and space began and will continue to be the rules until time, space and all the other fields of the NSU end.

This theory also implies that the black holes at the center of every galaxy in our universe are/is the same black hole. We are merely observing the contact points of our NSU with the SU. It is these contact points through which we observe the SU filling our "empty" space to create gas, stars, galaxies. It also implies that as "black holes" are part of the One (1), the black hole is conscious. So, what did happen? Imagine that you are the One (1) and that you desire to create something different from yourself, something with free will – a place with time, matter, and energy that was not subject to your will, to your infinite knowledge and understanding. You would need to create an NSU. A place with Uncertainty, time space, matter, and energy and the other differing fields of the weak, strong and magnetic forces, so that change, or as we describe it today, evolution, could operate, although in a finite manner – to create consciousness with free will.

Time and space, along with all the other fields, were created by the Singularity when it created the NSU. Observing all this from the NSU leaves us in a bit of a physics quandary. We know we are not infinite (entropy), we know we are subject to the laws of space and time, and we know we cannot travel faster than the speed of light. Observations of our Universe (NSU) indicate it is finite, not possessing enough matter to continuously expand and contract--although, even if it did contain sufficient matter, it still would not solve the space-time paradox because our finite physical environment makes it impossible for us to exist (mathematically) in a Universe, which is infinite.

Essentially how does finiteness then "appear" in infinity?

The Revelation does not stumble on this point. When the SU (The One (1), the Black Hole, God) decided to divide off a portion of its infiniteness and make it less dense (actually it's part of the SU, just time delayed), it did so by creating a 2 (the first particle of the NSU's beginnings). Again, this is the number 2 *and* the second "proton" physically. This is the moment the NSU begins. If you were the One (1)—and you were infinite and you choose to divide, the thing you created by that division, by mathematical law, would have to exist outside of infinity since anything divided from infinity leaves infinity as a remainder--but cannot be infinite itself. You can't have two infinities—by definition. Math doesn't create this seeming paradox of course -- *it solves it.*

Since the One (1) would have existed without the 2 prior to the 2's creation, the 2 must exist *after* the division. This "after" is a new requirement for the material formerly from the SU, *time*. It also necessitates the creation of *Space* at the boundary--a place contained within the One (1) but different from it. Space is a home for the 2 outside of the Singularity (no time or space in the SU). *The 2nd proton is made up of the material of the One, but separate from it in space and time from the NSU's point of view.*

According to the laws of mathematics this would make the NSU a "new" place on the other side of a boundary. The boundary between them is where time stands still, what physicists call the event horizon. Only the creation of our universe in this order, can logically and mathematically explain such a place. This is the interface between the SU and the NSU. Light is also an interface between the SU and NSU and is also a boundary. Since the 2 would have to exist *after* the One (1), it would be subject to properties that separate it from the One (1): by order of creation (came *after*) -- time; and by space (*outside*) the One (1). This creates the three dimensions of space (3) and one dimension of time (1) simultaneously. They are firmly linked together which leads us to a better understanding of how acceleration in space affects time. Light acts like a wave on one side of this boundary and a particle on the other.

It Is All a Matter of Time

The division process itself would require the separation of the NSU from the SU at a specific velocity (as viewed from the NSU space/time observation point), which we know to be 186,000 miles per second – the speed of light. Since all matter, time and space come from this division, *no matter (or energy) existing after the division can go faster than this velocity without going back in time by leaping ahead of the matter and energy divided out of the One (1) to a place before time and space existed – the SU.* Since exceeding 186,000 miles per second would bring the super-light speed matter back to the Singularity, it would have to violate a number of rules discussed above provided it was matter from the NSU, had mass, and started out going slower than the speed of light.

At the interface then, (the event horizon), we get the creation of our space and time. They are both mathematically necessary to accommodate the division/separation process from the Singularity. We know from Einstein's equations and subsequent experiments that mass and energy are just different states of the same material; literally Mass = Energy multiplied by the speed of light squared. Things in the NSU can change mass and velocity through the application/subtraction of energy. As they increase in speed inside the NSU, they increase in mass, and time will appear to slow down—all consistent with Relativity. The observed universe is consistent with these principles.

Time is the velocity associated with the creation/division/separation of space for a separate NSU inside the Singularity, *the* black hole, God. Light is observed by us to travel at the speed of this division, and therefore is our NSU's perceived speed limit—because we observe this event from the NSU side of the equation. As derivatives of all matter created in our NSU by the SU, we *cannot observe anything traveling faster than the speed of light* because the NSU we live in was created at that velocity. To go faster would entail being part of the SU, an alternative point of observation not available to anyone in the NSU. We can never observe objects traveling faster than light speed because we were created through processes that defined limits. The SU is infinite, not us. The SU is a singular point of view (one variable universe) and we are conscious creatures created in a two-variable universe. No matter how certain something is in the NSU, the Uncertainty Principle demands at the end of the equation that a second variable, another solution, an observation from a different point of view—be possible.

> The creation of an alternative point of observation, a second variable, is the mathematical necessity behind the Uncertainty Principle in the NSU, and is necessary for an SU to create an NSU containing free will.

In other words, the Singularity has one point of observation. To create a new universe that is not the "One (1)," a second point of observation must be possible. This necessitates the Uncertainty Principle and makes free will possible. The creation of the second variable is tied to all these concepts. We cannot experience the Singularity's point of observation, which is why our minds cannot completely understand infinity.

Infinity is a one variable concept. Our minds, like our NSU, are based upon two variables. This leads to another reason we can't physically see a black hole; they aren't in the two-variable universe. We are constructed to perceive only that which is in our universe. All these phenomena are tied to the same moment of division, the same rules, the same math, and the same process that created us and will eventually re-absorb us.

The Singularity appears in the Revelation as a One (1). It has no space or time and is of infinite density because it is a one variable universe. Since the two universes intersect, we can observe certain aspects of the Singularity from our Non-Singularity point of view. We observe certain properties of the NSU material (fields) literally change as they approach the SU. Some appear to defy the rules of the NSU all together; essentially flagging them for us as things, which either represent the intersection point of the two universes or underlying principles of the SU.

The following fields demonstrate this concept better than I can describe it.

All matter (a field) is subject to the Uncertainty principle as is all energy (a field) except photons. Photons are certain, never varying in intensity, size and only come in quanta, forming the cosmic definition of that word. So when dealing with matter or energy (two different presentations of the same material), you must deal with Uncertainty until you reach the place where the SU and the NSU intersect, which is the photon; and, suddenly there is a unique certainty.

All matter and all energy (in fact, all the fields in the NSU) are subject to space and time except gravity. The effects of a change in any one "place" within the time-space continuum must affect the rest of the continuum on a delayed basis relative to their acceleration *except for* a change in mass that would simultaneously affect the rest of the continuum's gravitational behavior. Gravity does not obey the speed limit or "rules" of the rest of the NSU materials (or fields) because it is a direct connection to the Singularity. When the Singularity divided, when it created the less dense NSU, it could not *completely* separate from the NSU. The addition of the second variable (free will) did not preclude NSU material remaining connected to SU material by an elasticity that we observe/describe as gravity.
As the material of the NSU itself approaches the speed of light, it increases in mass and slows down in time, demonstrating yet another edge of our universe and the place where the SU begins. We cannot reach absolute zero in the NSU for the same reason; this is another "edge" that has the SU on the other side.

All of the above seeming contradictions are provable. Many of these "unexplainable events" have been observed and are logically necessary when material from our NSU has its different fields (weak and strong nuclear force, matter, magnetism, gravity, etc.) fused back together to create a material of infinite density with no motion, temperature, and experiencing no time. (E.g. why photons appear to stop at temperatures that approach absolute zero.)

Pi is infinite **and** random; nothing else in the NSU is. Prime numbers are also infinite but not random. According to the Revelation, both of these mathematical abstracts form part of our interface with the SU.

In the Revelation, a stream of material exits the poles of a spinning (from an NSU point of view) black hole. The exiting material arcs toward the axis of the black hole, (the interface or the event horizon), creating matter and anti-matter at the point that they meet. The matter remains in the NSU as the anti-matter is absorbed back into the SU, mirroring its time separated twin. The matter becomes the hydrogen building blocks of our NSU. This is the place in the Revelation, where I admittedly lack the knowledge to completely explain the rest of the images in the Revelation. As a disclaimer here, I would like to say that I can, and have attempted to, safely describe 99% of the images in the first section of the Revelation, which is all contained in Book I (Peace Machine Hypothesis) of this work. I can safely describe 95% of the images in Book II (Market Fences and Economics) of this work but can only safely describe about 50% of the images in Book III (Unified Field Theory). Again, if this shakes your confidence in this work, I understand, but I think the reader needs to know this.

We Exist and Are Not In The SU – The Philosophy Basics

At the moment of creation when time and space began to apply to the newly divided particles, the NSU became the 2nd number. The only alternative to the creation of the 2 (or the 2nd), is for the NSU not to actually exist. I reject this idea for several reasons, all of which involve human consciousness.

All humans know is we exist. "Cogito ergo Sum", or "I think, therefore I am," a message from a mathematician – Rene Descartes. Even if we are a piece of software in a fictional program, that thing must also exist. This reduces the possibility that we don't physically exist to impossibility. The chance that we don't exist here in the form we think we do is also reduced to a very small probability, greater than zero but not quantifiable.

So, let's accept that we exist, and probably exist as we think we do. We believe we are the sum of many small parts. At a minimum, we are a part of something, a positive existence, something greater than nothing. By definition

The Revelation

anything greater than nothing must be at least a one. Here the circle actually starts again, but in a different universe, the one we are in, the one subject to time and space...the Non-Singular Universe or NSU. Our Universe, which we will now define as everything we *can* observe, must therefore be at least a 1 or greater, or it cannot exist at all. Since we exist in it, the NSU must have started out as ≥ 2 (the first prime number in our definition of primes) and must equal something finite--according to our definition of the One (1) and the mathematics of division. We can deduce this since we know that we:

1. Are not the original "One (1)."
2. Are *not* infinite.
3. Are not amorphous.
4. Are not a singularity—This last point demands the expulsion of a 2 (2nd) from the One (1), forming the NSU through a Big Bang event *or an alternative explanation*. Mathematically, we know the SU may contain charge because -1 times -1 = 1. (I believe this "negative" *expression* in the Singularity takes the form of charge.) In other words, negatives can become positives because they represent charge or a state of matter and energy. This implies anti-matter and anti-energy, which the NSU's time and space *cannot constrain,* as they are still part of the "One (1)" and are not limited by the division "boundaries" that created the NSU.

Fractions:

The NSU must also be greater than 0. Fractions are not allowed when only the One (1) exists. Fractions only exist when the One (1) and the 2nd (and all its derivatives) co-exist. Something cannot exist which is a fraction of something greater than is possible. Any such fraction would become the One (1) itself by mathematic law and by logic. Therefore any fraction of existence would by definition be one universe. In other words, an indivisible piece like ½ of one, where the whole One (1) doesn't exist—cannot be.

So, we exist. And we know we are not the One (1), which is infinite and not subject to time and space. If we were the One (1) and we were conscious, we would somehow *know* we were the One (1st) for many reasons, some of which are listed above. And, if we were the One (1st) and were not aware of it, time would not be affecting us, and apparently it is. So we are a derivative of the 2nd proton/universe after being divided out of the One (1). The universe as we know it, minus black holes, which we cannot see and are not part of our universe, could have been formed by the resulting explosion (expansion) from the first meeting of the 2 ($2^{nd\ proton}$) and its derivatives, as it was divided from the One (1) at the start of time and space. This would hold with current cosmology which supports a Big Bang...and which our theory doesn't rule out or rule in. Background radiation observable by instrumentation requires me to include for clarity, no matter how slight the chances, a *possible* Big Bang component in the origin of the NSU.

Spiral Arms Theory of Creation – An Alternative to The Big Bang

We theorize there was no "Big Bang." I, along with Dr. Richards, offer another mathematically possible explanation (our Spiral Arm Theory), which has the SU emitting energy/matter into our NSU at the Interface. Remember the Interface is at the time event horizon of the black holes. Black Holes are simply creating gas clouds filled with stars along the rotational access of the spinning black hole (spinning from the NSU point of observation)—by creating matter which appears to be materializing out of the expanding vacuum (dark matter). This is also the conduit for dark energy radiating from the black hole(s).

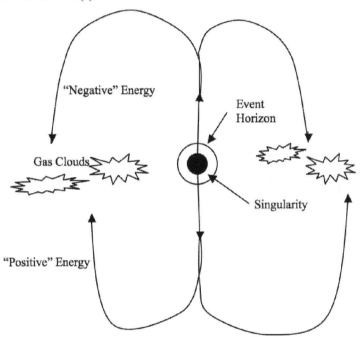

This matter would clump where space-time "kinks" around the gravitational fields of previously generated matter, which originally appeared randomly on this axis. The magnetic/gravitational forces of the BH might also concentrate the created matter on the equator of the BH. Gravity would quickly create a bigger "kink" until you have a cloud dense enough to form stars. This would create early galaxies with blue, hot, mostly hydrogen-comprised stars—again, consistent with astronomers' observations. These clouds then would trail out from the BH, forming arms, which of course would bend back in from the gravitational attraction of the BH, deep in space and time.

Scientists currently call this moment of separation the Big Bang (in one theory). The above explains why there could have been such an event at the beginning of time, without a Big Bang. Existing formulas get us from a nanosecond after the big bang to here—the question is what created the "instability" necessary in the early universe to take us from the infinite One, (1st) proton to the creation of the 2 (2nd proton). I think the very concept of this instability is inaccurate. You don't need instability if the creation of the NSU was done be design, by a consciousness higher than ours. The idea of creation being a conscious act creates a new concept of change. The first change was done by something that had to create a method, a math and an alternative Universe, to enable the possibility of change (or free will when viewed from the NSU) in the SU itself. In other words, this demands that we understand a new definition of the word "Evolution" and how that process would look mathematically – from the observation point of a nearly perfectly compressed, infinitely energetic sphere. The math suggests it *had* to be a conscious choice, not just an "instability," or error. The One must have (1) decided to divide; it "wanted" to be able to change, to evolve. Therefore, this division (or separation) is *both* a mathematical operator, as well as a logical choice. **Evolution**, rather than being what Webster's defines as:

> "A gradual process in which something changes into a different and usually more complex or better form."

…is actually **the mathematical and physical ability of matter to change**. Change requires a two variable Universe. For something to change, there must be another option available, a second vector, distinct and different from the first vector. The One (1st) proton (black hole, God) was not subject to time before the split, and it can never be subject to it now or ever—by definition. This is also consistent with the observable universe.

Older Galaxies are Bigger!

We believe that the energy surging out of the poles of a spinning black hole is responsible for the creation of all the hydrogen gas clouds in the universe. This is the basis for our Spiral Arm Theory alternative to the Big Bang theory. This, for example, would explain why older galaxies are so much larger than younger ones. Since dark energy is the only thing that can leave a black hole after the original division, it crosses the interface, colliding with positive energy, creating matter and anti-matter. The anti-matter (not confined by our limitations of space and time) returns to the SU, **mirroring the positive matters every move atomically/quantum-ly.**

<u>Under this explanation, black holes don't "eat" or consume matter, they create it.</u> And they've been doing it since the beginning of the space-time

continuum. This matter, by prediction, would spontaneously appear in space's vacuum (Einstein's cosmic constant explained) in a pattern dictated by gravity. It would look like a flat donut around the black hole center, radiating in spiral arms because of the distorting effect of gravity on space-time. Once initial (randomly distributed) gas groupings occur, they further distort time around them bending the time/space continuum toward them from all sides.

Since the matter is literally appearing out of a vacuum (dark matter), it would do this more where the time space bundles are closer together, where matter had spontaneously formed first. This would rapidly become a self-perpetuating cycle, along a circular track with arms radiating out from the center along the axis of the spin. Our Spiral Arm galaxies would look exactly as Galaxies actually look today from our observation point.

All this gravitational mass would "slow" down time (or the initial velocity as it should be labeled) in the area of the space continuum where it is the densest, and allow it to speed along at "full light" speed in the areas in between--causing an observer in the NSU to perceive an expansion (Hubble). I believe this perceived expansion should be *accelerating,* although this has not yet been demonstrated.

Gravity

Gravity is the connection between the One (1) and the NSU. That is why it doesn't have a speed! When the One (1) expelled the 2 (2nd), it did so by converting some of the amorphous material of the One (1) into matter (and all the other fields like energy, time, etc.) as we perceive it in the NSU. The energy causing the separation and transformation of an SU component from an infinite density to the expanded matter composing the NSU must result in a counter force that continually seeks contraction to its original state: an anti-separation/expansion force so to speak. By the expenditure of energy, this force of Gravity can be delayed; *but it cannot be delayed forever.* Again, the combined laws of thermodynamics support this truth. Even matter converted in other manners, (fusion, etc.) will eventually, by nuclear decay or re-absorption, re-enter the SU.

In our NSU, gravitational forces increase the probability of matter being creating around other matter due to the curvature of space inward around mass. Scientists currently explain older galaxies being larger than young ones with a combination theory: old galaxies according to astronomers and scientists are the combination of multiple galaxies that have collided. *This is a ridiculous and an embarrassing explanation.* Astronomers should have just said, "Our observations don't match our theory, yet." The poorly-conceived collision theory creates an undeniable paradox *and* an observational conflict.

The existing, older galaxies *don't look like multiple galaxies*, which collided (statistically they would collide from every angle, meaning spiral arms juxtaposed, not layered). We have only seen a few example of this, not the millions we should see if this explanation were true.

The paradox in the collision theory comes from the conflict with Hubble's red shift, which makes this theory literally impossible. Draw 100 dots on a piece or rubber and then stretch it in every direction—how many of the dots connected up with other dots? None. In fact, every dot will be farther away from every other dot on the sheet. How would *every* older galaxy get bigger by this nearly impossible process? The real answer is that older galaxies are more massive because they have more material in them from another source. Since the stars within the Galaxy are converting some of that material into energy (E=mc2), we know for a fact that there is a separate process adding material faster than the stars are converting it into energy. Otherwise, the older galaxies would have less and less material, and should become *smaller,* as they age. Current cosmic theory adds to the total loss of matter (accelerating proposed galactic shrinking) by the assumption that matter is constantly being consumed in the black holes at the center of every galaxy! Black Holes have been *discovered* at the center of every galaxy. Older galaxies should therefore be shrinking most, and this is simply not true according to observation.

The Mass of the Non-Singularity Universe is Increasing

Older galaxies are bigger because they are growing, not combining. As described above, black holes are feeding matter into each galaxy, usually at a rate faster than it is being converted to energy or lost to space/entropy. If black holes were actually consuming matter instead of being responsible for its creation, the universe would look opposite from the one we observe.

The Unified Field Theory as presented in this book implies that we do not live in a steady state universe. Black Holes are actually funding the universe's matter and energy diet through the constant "creation" of energy and matter in the form of hydrogen creation out of dark energy. This is not to say that the SU is actually losing mass or energy, since when you take/divide something from infinity you are still leaving infinity. Still, this doesn't completely express what really happens. The energy and matter in the NSU are less dense components of what exists in the SU.

As the CLOT laws are truly observed everywhere, all matter and energy will eventually return to the SU. One way to test our theory as presented in this book would be to measure whether the mass per unit volume in space measured in multiple places is increasing. If all the measurements displayed an increase, something like Hubble's constant, it would lead us to a different cosmology and interpretation of the WMAP data than is currently being used today in the physics community.

The Time Limited Nature of Our Universe's Existence

Since the One (1st) proton created the 2nd and we know we are not the One (1), we must be the 2 or a derivative. This puts God/the Singularity/black holes/infinity/One (1) on the other side of the time divide from us, out of our observable universe. This position is completely consistent with Relativity and the observable NSU. If time stops at the edge of a black hole—we cannot travel to it. Time would infinitely slow down as we approached; and, NSU space/time parameters would never allow us to leave this Universe (the NSU), in our current slower than the speed of light state. We will enter the SU when the SU decides to reabsorb us--which only it can choose to do.

The SU is amorphous and accomplishes all this through the use of charge, or the use of different charges (the difference in charge doesn't contradict its amorphous nature). This charge presents in a two-variable universe as different states of matter and/or energy that are of the same value. We call four of these states, matter and anti-matter and energy and dark energy. Since all are part of the One (1), they are not subject to time and space while in the Singularity. We express these different states mathematically as charge, and have virtually no other information on them.

To divide the One (1) into a 2 (2nd), we need to divide the one (1) into *different things*. There would be a necessity to separate the amorphous One (1) into non-amorphous (us) material. Since truly independent material creation is mathematically impossible, and mathematics' rules are absolute --matter and energy must be the same thing but in different states. Einstein predicted this, with E=mc squared.

But, a mathematical expression of their different properties is permitted and observed. This expression begins at the interface, the beginning of time and space. Matter and energy with a conversion factor would work nicely—but matter and energy with different charges works even better, because this gives the NSU and the SU each a state of being with their own charge. Now we have a 2 (us), and this requires the introduction of time into the formula. Since the One (1) was infinite, it still must be. Therefore, the One (SU) must transition from being One (1), to a divided state and then back again.

This makes space, time, and matter all finite. Time began when the One (1st) divided, and is in existence only as long as the 2nd proton is still separate from the One (1). This is why time stands still at the edge of a black hole, the interface between the One (1st) or SU and the 2nd proton/space/time/NSU. The

2nd proton must begin after the One, and therefore must be divisible from and by 1 as is everything else from that point. The formula is written in the following manner.

NSU $\geq$ 2: Therefore the NSU must really equal 2 or more. Again, fractions don't work here for the same reason; the instant a piece of the universe separates from the first piece—it automatically becomes a whole number≥ 2, or the first real prime number in the NSU.

Prime Numbers

Why are prime numbers so important? Because they represent the method the SU used to create the NSU. The number system in the SU only contains one number; and the number system in the NSU only contains prime numbers. Humans have developed a number system that fills in the spaces between primes with other numbers. This is why prime numbers have unique abilities in all math(s). The beginning of the NSU was a prime number festival because it doesn't make sense to God to combine incredibly dense particles in a compressed state by anything except prime numbers. Any core multiple except 1 and the number itself would already be occupied by a combination of *other* particles more efficiently combined. So, what are the other numbers?

A four is actually two combined 2's—expressed much more efficiently than a new 4 would be, just after a 3 was created, but before a 5 was. Hence any numbers other than primes are really place holders in our mind (NSU) and *do not actually exist*. According to the Revelation, numbers themselves are not completely accurate; a 2 is actually not the sum of two ones, because those number ones cannot be described exactly. I am not going to expand on that in this book because I do not fully understand it. It's a part of the Revelation too complicated for me to comprehend. Biology is also driven at the base level of protein pairing by the juxtaposition of the combined laws of thermodynamics in what I call *the rule of singularity compression*.

This is the part of the Revelation that "explains" why prime numbers are very important in all mathematics. Primes represent the points where math scales change because their ability to get larger or smaller makes other combinations of previously existing numbers more likely from a statistical point of view than creating a new number. There can only be two divisors in a universe of a dual nature (the NSU); and only one on the Singularity side where time and space have no existence. The presence of two divisors, a product of the first division (from One (1) to (1) and 2), is integral to the NSU and all its math forms. The second divisor, the second variable, the second point of observation, the existence of free will, all begins here. These things are properties of the 2 (2nd) which comprises the Non-Singularity Universe

(NSU), a place where Certainty and Uncertainty both must co-exist within Space and the velocity of the division--what we call time.

Singularity material can exhibit to the NSU observer as either energy or matter, in a positive or negative state. This fundamental material/energy translates through the Singularity into our NSU in all its expressions (black holes, fusion, love, supernova) at the conversion rate of E=mc squared -- in both directions of the equation.

"Quantum Mechanics" is quite a large error in labeling; *it should just be called the Uncertainty Principle because it applies everywhere, even in the Singularity*. If it did not apply in the SU—that would lead to a *certain* description of the SU, making it finite...which by definition is impossible for the reasons explained earlier. The only other alternative is that we don't exist.

The conversion rate, since it is neither matter nor energy, can be constant and exactly known, hence bracketing the Uncertainty Principle. But, you can never calculate exactly how much energy is converted to matter or how much matter to energy. All we can know with certainty is the conversion rate at which it happens. Hence, the rate of conversion can be certain while the exact product output remains uncertain -- again obeying the Uncertainty Principle *and* our observations...but quite vexing for Mr. Einstein.

All observations in the NSU, by definition, are limited to a range of answers that exist within the permitted mathematical extremes of our observational viewpoint. Of course you can observe the moon, Mr. Einstein, it is in the non-singularity universe where observation by NSU occupants is certain and predictable. But, you should not expect to observe the Singularity, the One (1), from a point of view in the NSU, as we exist only as a *later* creation of the One (1). This should not have surprised Einstein.

The conversion of matter to energy *or* energy to matter results in an additional phenomenon: a change in the electron's oscillation rate. This change is environmentally triggered from the moment of the electron's creation. The energy transfer in the electrons' shells forms one of the NSU's connections to the SU. For us, this conversion of the electron into something that isn't "completely observably" is possible because part of the electron's matter is converted to an energy state that is neither defined nor limited by our NSU space and time confines. This occurs so that both the E and M products of the conversions remain *connected* to the SU. The Uncertainty Principe allows and demands this to be so.

How exactly is this done? Imagine that since an Electron oscillates at the speed of light, the energy level is already on full at all times. The speed limit of light makes this possible. Indeed, without such a limit, energy could be added, and no mirroring would be required in the SU. You cannot do anything in an electron's environment, without changing the energy level, without changing the electron itself. Again, you can't do anything without changing that oscillation. According to the Revelation, the Electron is "connected" to the strong nuclear force at the center of the atom and is oscillating at the speed of light. The Electron, by design, cannot absorb any of these external energy changes. Instead, the entire atom must change, move, and evolve to accommodate the new energy present. Since this change is then reflected in the relationship between the Strong Nuclear Force and the Electron, the Strong Nuclear Force at the center of the atom transmits through quantum mirroring that energy change to the SU. Since the reactions just described represent everything the atom can do, quantum mirroring then is completed through this process.

In fact, the photon is the only certain, non-varying piece of energy/matter in the NSU, and is the only thing in our universe not subject to the Uncertainty Principle. Every photon is exactly the same. To say God is Light, Certainty, and the Singularity actually makes literal sense. To have a universe of Uncertainty all end at one certain spot creates structure for our side of the universe. The photon provides that structure. Without Certainty here at the photon, all math, all matter, and energy, fail.

Review of Primes

If all primes were certain without exception, it would be a perfect description of just one point of view. In this way, we can see that to create a different point of observation, we had to start with a certain 1, (which of course has only one divisor, itself) and then progress to an uncertain 2 (has two divisors). The SU is a One whose matter/energy is perfectly compressed and amorphous and infinite. A piece of this initiated creation of our Two (the NSU at inception), and was an efficient expression of a *nearly perfectly* compressed material, except that it was a finite, imperfect sphere (as Pi is a non-exactly defined number), and the matter/energy of the Two (NSU) was *now* subject to space and time, also known as the velocity of the division @186,000 miles per second. Light energy got its waveform along the event horizon at the point of conversion of matter and energy that occurred when the SU divided to create the NSU.

After a 5, the combinations of 2's and/or 3's are more efficient than a six. They negate the possibility of a six. But a 7 is different. A five plus a two, is the outcome of these rules, which created the 2, 3 and 5 first, but left 4 and 6 as "holes" where other more perfectly compressed materials in combination are more efficient. For more on this topic please read Appendix I: More on Dr.

Hawking.

I don't completely understand this, but the way it appears in the Revelation offers a clue. The ratio of the surface area to total volume of two combined spheres that total a prime number are on one side of a dividing line, while the ratio of surface area to sphere volume that add up to non-primes are on the other side.

For the One to divide and create the duality which *we can observe with our consciousness as well as predict from this formula* at the center of the entire Universe--it had to choose to do this—and this choice implies consciousness! It also had to figure out a *way* to split. I believe it designed a mathematical model based upon the realities of being One (1) and physically becoming/dividing part of itself into a 2 (2nd) as described above. This makes God, The One, part of everything derived from that split. Scientifically you would say all matter and energy derived from the division of the One (1) would contain elements of its origin. The states of being which exhibit differently like energy and matter -- *must* be mathematically equal. They must, therefore, be some part of the amorphous whole from whence they came. According to the Revelation I experienced, at least one of those forces/material/fields is love because it specifically contained the formula: God is, (or equals, =) Love.

I think if we study the most efficient method for pushing a 2 (2nd) out of the One (1)—we will see why Pi cannot be calculated and is random. The edge of the interface must be "open" (not exactly definable) for this event to happen. Once infinity is perfectly described, nothing could change that description. No door out of a perfect description, no change, no second variable, no NSU. This establishes the Uncertainty Principle as necessary for the One (1) to create the NSU. Since the One (1) by definition is infinity *and* uncertainty (indescribable), then the 2 (2nd) cannot be completely certain or uncertain either. That means the Universe we live in must contain uncertainty when attempting to observe (and define or describe) all Quantum or SU adjacent elements. Again, this is consistent with the Copenhagen Interpretation, and our observation of the Universe.

Why God Is Love

The first sphere, the infinite One (1) or God, created the Two upon *deciding* to divide off a portion of itself. This purposeful act made it possible for everything else we observe to exist in the NSU. **If The One chose not to separate, we would not exist.** It used, and is using, a lot of energy to create us and hold our Universe open in our less dense state, defying gravity for now.

Addendum: A View of the Cosmic Time Hypothesis (CTH) and the Spain/Richards Unified Field Theory (SRUFT)

In the final phases of editing this book, several months prior to publication, Dr. Richards came across a very interesting theory called the *Cosmic Time Hypothesis* (CTH). It was found by a search on Google for "cosmic constant." CTH appears to have originally been postulated by Horst Fritsch in April of 2000, according to the signature at the bottom of the web page. We could find no other references to this hypothesis. Nonetheless, we found it fascinating because CTH looks at time as being different when viewed from our point of view versus the cosmic point of view – We would say a NSU versus SU point of view. While there are many similarities in the conclusions, there is no logical derivation offered by Mr. Fritsch for his theory. For completeness I have copied the Cosmic Time Hypothesis below, followed by our comparison of it to the SRUFT.

Cosmic Time Hypothesis -- A Summary

Cosmic Time Hypothesis (CTH) suggests a new cosmic model of the world. It is supported by three well-accepted axioms:

1. The velocity of light is a universal, natural constant.
2. The general theory of relativity *(GTR)* is valid for the whole universe.
3. Averaged over great distances, the universe opens up a flat (Euclidean) space.

Based on this: *CTH* results from the field equations of *GTR* when abstract Newtonian time is replaced by the Cosmic Time, its rhythm being determined by the expansion velocity of the universe.

The *CTH* convinces not only by its internal consistency and simplicity, it transmits also surprising perceptions with far-reaching consequences for the physical conception of the world and especially for cosmology and the physics of *elementary particles.*

The most important results of *CTH* are:

- There is a far-reaching agreement with the standard model of cosmology - big-bang-theory - if Newton-Einstein-Time t is used as time parameter. *CTH* elegantly solves the main problems of big-bang-theory (problem of the horizon, the formation of the galaxies and the flatness) without using the inflation model.

- Differing from big-bang-theory, the following also results from *CTH*:
  - The gravitational constant *G* is not constant. Instead, it decreases versus time: $G \sim t^{-2/3}$.
  - Always and everywhere in the universe matter is generated continuously: $M \sim t^{2/3}$, $GM = constant$
  - The universe expands with the velocity of light:

$$\frac{dR}{dt} = c \sim t^{-1/3}$$

Velocity of light *c* always is measured as a constant

$$c \, \Delta t = constant; \quad \Delta t \sim t^{1/3}$$

- *CTH* postulates a cosmic time $\mathcal{T}$, which is connected to Newtonian Time *t* with the same transformation equations

$$\mathcal{T} \sim t^{2/3}; \frac{\Delta \mathcal{T}}{\Delta t} \sim t^{-1/3}$$

as the real "inertial systems" (being in a gravitational correlation) with the idealized inertial systems (free of forces) of Newtonian and Einstein's physics.

- The basic terms our physical world: Space (*R*), Matter (*M*) and Time ($\mathcal{T}$) are interconnected by the *CTH* in the most simple way conceivable: $R \sim M \sim \mathcal{T} \sim t^{2/3}$

- The Cosmic Time shifts big bang to an infinitely remote past. According to this, the universe, measured in its "Own Time", would be infinitely old:

$$\frac{\Delta \mathcal{T}}{\Delta t} \sim t^{-1/3}$$

- Real clocks (pendulum and atomic clocks) do not measure Newton-Einstein- time *t* but instead cosmic time $\mathcal{T}$.

- *CTH* agrees with the so-called Super-Symmetry-Theory which postulates that at Planck time ($t_{PL} \approx 10^{-43}$ s after the big bang) all natural forces were identical. Following *CTH*, gravitational force at Planck time was as big and had the same radius of influence as the strong nuclear force today. As long as the universe has existed, it has obeyed the preservation axiom: "The product of the reach of the gravitational force and its strength is constant and always equal to the strong nuclear force."

-
- *CTH* solves the problem of the cosmic constant. It explains why between the theoretical energy density of vacuum resulting from physics of nuclear particles and tied to the cosmic constant Λ and its observation a discrepancy of 120 orders of magnitude opens up. For some well-known physicists this discrepancy presents a basic problem.

Following *CTH* the cosmic constant is not the reason for a "fifth natural force" accelerating the expansion of the universe. Contrary to this, it supplies - like matter and radiation - a contribution to the total gravitation field of the universe. Also its value is not constant.

$$\Lambda = 1/4R^2 \sim t^{-4/3}$$

Instead, it decreases versus time. Due to that, the cosmic constant - introduced by Einstein and later discarded by him - does not exist.

- The *CTH* harmonizes "Big-Numbers-Theory" of Paul Dirac, where the gravitational constant versus time is decreasing. It so also supports the "theory of expansion of the Earth" developed by Pascual Jordan, which is an elegant and convincing explanation for a number of geophysical occurrences.
- The *CTH* explains in a simple and natural manner some phenomena of observational astronomy:

1. The flat space-time existing up to the visible borders of the universe.

2. The visible and continuous generation of new stars and galaxies everywhere in the cosmos $M \sim t^{2/3}$.

3. The fact that we observe less matter in the universe as it would be necessary for a flat space-time $M \sim t^{2/3}$.

4. The decrease of the rate of generation of new stars $\frac{dM}{dt} \sim t^{-1/3}$, derived from observations.

5. The discrepancy between distance and red shift of type Ia-Supernovae in remote galaxies, resulting from measurements.

- The distinctive feature of *CTH* is the fact that it displays two faces. One face is revealed if one views *CTH* through "*t*-time-spectacles". The other (true) face can only be seen through spectacles of $\mathcal{T}$-time. If this is done, it leads to the paradox situation, that big-bang-theory in its leading characteristics matches with *CTH* (t-time) as well as steady-state-theory ($\mathcal{T}$-time), once being a fierce competitor of big-bang-theory.

Horst Fritsch
2000-04-09

The Cosmic Time Hypothesis (CTH)
Vs.
The Spain/Richards Unified Field Theory (SRUFT)

1. CTH suggests support from three axioms:

   A. The velocity of light is a universal constant. SRUFT agrees with this, proposing that this constant is the result of the process by which the NSU is created. In other words, SRUFT proposes that the division of the 2nd particle from the SU is done at the speed of light. This is consistent with the CTH proposition that Cosmic Time (CT) is:

   "…being determined by the expansion velocity of the universe."

   B. General Theory of Relativity (GTR) is valid for the whole universe. Fritsch is saying we don't need to differentiate between Special and General relativity because General Relativity would work even in an inertia system. In the SRUFT, the SU is an inertia system, not idealized, but actual. For this reason, we, the authors of SRUFT, agree completely with this assertion. Reconciliation of the GTR with Uncertainty (Quantum) through the creation of a real inertia system is central to our description.

C. Averaged over great distances, the universe opens up a flat (Euclidean) space. SRUFT agrees with this conclusion but not the process to achieve it. CTH assumes a big bang with this disposition while SRUFT assumes a steady increase in matter produced at the SU/NSU Interface (or event horizon), from a fast spinning sphere of nearly perfectly compressed material (our SRUFT inertia system: SU, black hole, etc.) which then creates a flat (Euclidean) universe. This happens because the point of creation is only a point, a tiny fraction of the whole, so the clouds of matter (hydrogen) "bend" space-time, making it increasingly likely for an "axis" of matter to form. CTH does admit the mathematical possibility of a steady state. In an interesting paradox, we propose a steady-state version of the universe, but our math forces us to admit the possibility of the big bang. His math convinces him of the big bang but also forces him to admit the possibility of a steady-state version. Very unifying really.

2. According to Richards and Spain, the most important results of *CTH* by Fritsch are:

There is a far-reaching agreement with the standard model of cosmology big-bang-theory - if Newton-Einstein-Time (t) is used as time parameter. *CTH* elegantly solves the main problems of big-bang-theory (problem of the event horizon, the formation of the galaxies and the flatness) without using the inflation model.

Differing from big-bang-theory, the following also results from *CTH*:
0   The gravitational constant G is not constant.
1
2   Instead, it decreases versus time: Always and everywhere in the universe matter is generated continuously: $G\,M = constant$
3
4   The universe expands with the velocity of light: Velocity of light c always is measured by us as a constant.

These findings are also critical to SRUFT. Of particular agreement are the statements concerning the continuous generation of matter in the universe (which is fully supported in the SRUFT in several places) and the perceived speed of the expansion of the universe, (also explained and predicted in the SRUFT).

According to the CTH: "The Cosmic Time shifts Big Bang to an infinitely remote past. According to this, the universe, measured in its "Own Time", would be infinitely old." This, in the SRUFT, is the SU, (the inertia system) and indeed, our own description matches the CTH here.

According to the CTH: "Real clocks (pendulum and atomic clocks) do not measure Newton-Einstein- time t but instead cosmic time." This is *almost* accurate. Actually, according to the SRUFT, they measure everything from Planck time (10 to the negative 43 after the Big Bang or separation of the Two from the SU) to CT, which is *why they can disagree*.

According to the CTH: *CTH* agrees with the so-called Super-Symmetry-Theory, which postulates that at Planck time (10 to the negative 43 after the Big Bang) all natural forces were identical. Following *CTH*, gravitational force at Planck time was as big and had the same radius of influence as the strong nuclear force today. As long as the universe has existed, it has obeyed the preservation axiom: "The product of the reach of the gravitational force and its strength is constant and always equal to the strong nuclear force." SRUFT largely agrees with this CTH conclusion.

SRUFT requires that all natural forces were actually equal and shared the same radius of influence because all of them are components of the *complete* strong nuclear force (SU) before what we would call expansion/creation of a 2 or division. This is the moment of the creation of the Interface, or the reduction of density to create the NSU from nearly perfectly compressed material that comprises the SU, which is the original inertia coordinate set. This includes a quantum element, the NSU's share of the SU consciousness, as well as all the other natural forces.

According to the CTH, this solves some observation paradoxes with current cosmology like:

"The visible and continuous generation of new stars and galaxies everywhere in the cosmos."

SRUFT shows that this paradox not only disappears with this cosmology, but also more fully explains it. CTH also calls for the gravitational effect in the universe to decline as time progresses. To us, the most amazing part of the CTH cosmology is the author's explanation:

"The distinctive feature of *CTH* is the fact that it displays two faces. One face is revealed if one views *CTH* through "t-time-spectacles". The other (true) face can only be seen through spectacles of cosmic time. If this is done, it leads to the paradox situation, that big-bang-theory in its leading characteristics matches with *CTH* (t-time) as well as steady-state-theory (cosmic time), once being a fierce competitor of big-bang-theory."

SRUFT of course, agrees with the CTH in both of these two critical areas. The "first" and "second" face of time described in the CTH is in reality a view of describing time from the perspective of the SU and NSU -- in the SRUFT. This is fully explained in the SRUFT as necessary because it represents the second point of view, the second point of observation. The SRUFT describes a universe where time and space do not apply and one where they do, under cosmic time. This relationship is demanded in the SRUFT by the creation of time and space as a result of the division of the NSU from the SU. The second area of agreement is the possibility (explained in the SRUFT and CTH), of either a big bang cosmology or a type of steady state cosmology. We state the NSU is a steadily increasing mass universe, with the SU funding the expansion through expansion of SU material into NSU material (strong force components), appearing in our universe (time space Euclidean field) as it is fed from the SU at the speed of light.

The real metaphysical question remains. Since the SRUFT offers a complete explanation for the order of this process, the reasoning, and results that largely match this CTH equation, where did the Horst Fritsch get this hypothesis? Did he too have a Revelation?

Appendix I – More on Dr. Hawking

Dr. Hawking has correctly proposed in much of his writings how entropy is related to the combined laws of thermodynamics. But, he sees one as a result of the other. We believe that one doesn't cause the other -- they are the same thing. This 'causation" error also has the physics community describing the combined laws of thermodynamics as something which;

"...Does not hold always, just in the vast majority of cases."

Dr. Hawking describes a glass case with a partition separating oxygen and hydrogen molecules, the mixing of which would occur after the removal of the partition:

"The probability of all the gas molecules in our first box being found in one half of the box at a later time is many millions to one, but it can happen."

This cannot be so. The second law of Thermodynamics is *never* violated in our side of the universe. The atoms in Dr. Hawking's box would never return to their ordered state without an ordering mechanism and the expenditure of energy to operate it. These are important principles because they tie together how the Singularity divided from our universe, and how NSU matter behaves as a result of this. It is known, for example, the oscillatory motion of an electron increases as its relative speed to our matter decreases thus making the net velocity equal to the velocity of light. Again, this is the result of the combined laws of thermodynamics and is connected to the origin of the energy source driving the electron's amplitude. Electrons can pop out of the vacuum and have set quanta of energy because they are the manifestation of light (radiation) and energy at the interface between dark matter and regular or positive matter.

Quanta here are a conscious choice: there is no path from this side of the universe back to the Singularity except by the Singularity's choice, one of which appears to be a supernova. The photons quanta must be discreet because the effects of time and space exist in the NSU but not in the SU. Hence, SU material is snipped off into discreet quanta as it traverses the interface.

The transition of photons leaving the Singularity and entering our Universe (NSU) also necessitates the creation of the Uncertainty principle since they transit from high energy at no velocity to low energy at high velocity, literally becoming the connection at the interface of the SU and the NSU. We can't say precisely where they are because we cannot enter, describe, or find the edge of the Singularity, although it can send us dark energy, which changes state and becomes matter. This why an electron can resemble a wave, have mass, and still oscillate at the speed of light. Electrons carry one quanta of energy but cannot freely re-enter the Singularity since the electrons existence in the NSU represents an irreversible change of state.

Mr. Planck would have said the exception to the second law of thermodynamics would occur if the initial energy levels were at an absolute maximum or minimum, a theoretical state not currently observable (with good reason) in *our* universe. To understand both of these men, we have to go the source. I will quote Mr. Planck, since his description of the situation cannot be improved upon:

> "If, for instance, an exchange of heat by conduction takes place between two bodies of different temperature, the first law, or the principle of the conservation of energy, merely demands that the quantity of heat given out by the one body shall be equal to that taken up by the other. Whether the flow of heat, however, takes place from the colder to the hotter body or *vice versa*, cannot be answered by the energy principle alone. The very notion of temperature is alien to that principle, as can be seen from the fact that it yields no exact definition of temperature. Neither does the first law contain any statement with regard to the direction of the particular process.
>
> Means only that if hydrogen and oxygen combine under constant pressure to form water, the re-establishment of the initial temperature requires a certain amount of heat to be given up to surrounding bodies; and *vice versa*, that the amount of heat is absorbed when water is decomposed into hydrogen and oxygen. It offers no information, however, as to whether hydrogen and oxygen actually combine to form water, or water decomposes into hydrogen or oxygen, or whether such a process can take place at all in either direction. From the point of view of the first law, the initial and final states of any process are completely equivalent.
>
> In one particular case, however, the principle of conservation of energy does describe a certain direction to a process. This occurs when, in a system, one of the various forms of energy is at an absolute maximum, or absolute minimum. It is evident that, in this case, the direction of change must be such that the particular form of

energy will decrease or increase."

Dr. Hawking -- and many others to be fair -- believe for example that if you filled a cube with oxygen and nitrogen atoms segregated by type and then removed a partition between them, they would randomly mix, becoming "disordered," and that the entropy of the gas has gone *up*.

This is tied to Dr. Hawking's beliefs about black holes in which he states the entropy of the universe itself would go down, or become *more* ordered, if gas was absorbed or swallowed by a black hole without the use of energy. This cannot happen, because entropy never decreases in our universe. Things can become more orderly through the use of energy, but that simply has no significance to this discussion because to do that would require the use of energy without the progression of time (an impossibility). However, this process is possible, but in the Singularity (inside a black hole), which is one reason matter from here (our universe) cannot go *there* without being converted into a state or property of the Singularity itself, e.g. through a supernova collapse/explosion.

Many scientists will contradict this, saying a black hole is formed in the first place by the collapse of a giant star, taking a huge amount of mass with it into the Singularity. I don't dispute this. I think the Singularity can reclaim mass and energy from "our" universe *in this manner*, because, according to the Spain-Richards theory presented in this book, it radiates dark energy *out*, which is converted in the space-time continuum into an electron and a positron.

What is happening in a supernova is the outward explosion of most of the star's mass into the space/time universe and the conversion of some of its mass, and a lot of positive energy, into dark energy and dark matter. They convert at the instant of re-absorption into the Singularity. Since dark energy and dark matter do not have mass as we measure it, the time/space continuum rules need not apply. Black holes, then, are the door to the Singularity—and as predicted by this definition, part of the Singularity itself.

Dr. Hawking also believes the total area of the event horizon would increase by the combination of two black holes. The math, which led him into this area, is correct, but his conclusion was erroneous. The event horizon (the interface) of two combined black holes (nearly perfect spheres) would decrease proportionally to their volume were they to merge. Two spheres which can undergo no more compression (black holes for example), when combined, would have a total surface area proportionally *less* than the combined surface area (also a description of the event horizon or interface) of the two spheres measured separately.

Dr. Hawking's conclusion is true only if they inefficiently merge. I call this the inefficiency principle, which is a law unto itself in mathematics, although not yet discovered. So, a one and three cannot become a four, because this would be fundamentally inefficient in an almost perfectly compressed environment: two 2's would be more efficient. Indeed, in the Revelation's cosmology, the four does not exist at all. However, there is no more efficient way to combine a 2 and 3, both primes, than a 5, another prime, so this can be done and is. Primes and Pi form the basis of the efficiency principle, which in nature can never be violated at maximum compression. Since we don't exist at maximum compression (we are very inefficient, with fours everywhere) the number four exists for us.

Appendix II -- Sectional Cities

A sectional city is one way to revitalize our blighted urban areas. To reach its productivity potential, land use must have a density and tax level sufficient to support many CHUs in order to finance infrastructure growth, and necessary services. Many municipalities currently chase this dream in destructive ways. Free-trade zones, long-term tax breaks, even the construction of infrastructure for a specific business, are all very poor uses of public money which have been tried and failed. The ones that failed completely left everyone unhappy and the taxpayer less wealthy. The ones that "succeed" leave the developer wealthy and the taxpayer less wealthy, thinking the developer "made out" at his expense. Historically, this is almost always also true.

Unfortunately, the people running cities are unaware of the true laws of economics, and therefore do not understand that they already control the most critical elements for the success of business plans that depend on cities. Proximity is an economic principle just like other physical forces. It has effects like other forces. In nuclear science, one would call it critical mass, the point at which density changes the properties of the environment in a geometric rather than linear fashion. Proximity force in economics creates a corridor that runs through financial institutions, legal institutions, government and service institutions, and some business corporations that have major influence. Traditionally, cities attempt to tax everything they can around these corridors. Politicians have learned from experience that the Proximity environment will bear this load, making cities less and less productive over time.

If just 50% of the tax revenue from the bridges and tunnels around NYC had been re-invested, it would be a transportation marvel instead of the disaster it is today. Proximity to these corridors is measured in feet, view, parking privileges, and class of space -- all of which are combined by the laws of supply and demand into price or rent. Property taxes adjust with property values and act as a way for the government to receive a proportional share of this inherent value as it goes up and down. This is not meant to be a complete attack on the concept behind property taxes because the sectional city concept relies on a type of property tax.

Sectional city politicians would create special business zones. These zones would be designed around the Proximity corridor, and "line up" geographically with the next closet major city's proximity zone. The infrastructure base, including a transportation system (moving sidewalks would be good), would be constructed by the city along this vector. Finished construction would be allowed on the accompanying infrastructure platform next to the transportation resource. Renewable but not permanent items, like

99-year leases. In reality you would create a special district, seized by eminent domain and reconstructed with a transportation system that moved people but not cars. Cities would then rent the space along the transportation corridor for premium rents. This could be the first step toward a new definition of public and private space that mankind needs to tackle urgently for economic reasons. See the drawings below. The idea would be to have the central corridors serviced by truck, train, and auto traffic from the backside of the transportation corridor. Once a person steps off the platform onto the walkway and begins moving forward by walking, he catches the second belt and can move onto it as easily as he moved on to the first belt. This is relativity. The same would be true as he continued to move from belt to belt, picking up 3 or 3.5 mph this way, per belt. I think 18mph is practical this way.

Outer lane walkers could do 18-20 mph!

Platform

Lane 1, traveling 3 mph

Lane 2, traveling 6 mph…and so forth, add walking speed of course.

With a two or three mile long city core, this would move people from end to end, a lifetime of experience in a major city, in 8 to 10 minutes. Since we only moved the human body and the accompanying walkway material (no car and engine mass), we save a huge amount of energy. We also create a healthy walking environment for humans. We were walkers, for millions of years.

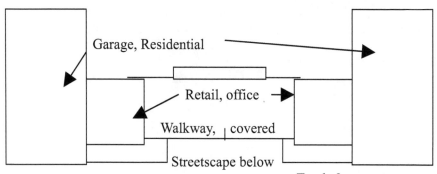

God is love

Appendix III, The Itch

I have shortened this article considerably to fit it in. In fact, I cut out the part about itching completely, which is why the articles title may not make sense. Read it, it's worth the time. The original appeared in the June 30, 2008 edition of *The New Yorker*, by Atul Gawande, a truly great writer and researcher.

THE ITCH

Its mysterious power may be a clue to a new theory about brains and bodies.

by Atul Gawande

A new scientific understanding of perception has emerged in the past few decades, and it has overturned classical, centuries-long beliefs about how our brains work—though it has apparently not penetrated the medical world yet. The old understanding of perception is what neuroscientists call "the naïve view," and it is the view that most people, in or out of medicine, still have. We're inclined to think that people normally perceive things in the world directly. We believe that the hardness of a rock, the coldness of an ice cube, the itchiness of a sweater are picked up by our nerve endings, transmitted through the spinal cord like a message through a wire, and decoded by the brain.

In a 1710 "Treatise Concerning the Principles of Human Knowledge," the Irish philosopher George Berkeley objected to this view. We do not know the world of objects, he argued; we know only our mental ideas of objects. "Light and colours, heat and cold, extension and figures—in a word, the things we see and feel—what are they but so many sensations, notions, ideas?" Indeed, he concluded, the objects of the world are likely just inventions of the mind, put in there by God. To which Samuel Johnson famously responded by kicking a large stone and declaring, "I refute it *thus!*"

Still, Berkeley had recognized some serious flaws in the direct-perception theory—in the notion that when we see, hear, or feel we are just taking in the sights, sounds, and textures of the world. For one thing, it cannot explain how we experience things that seem physically real but aren't: sensations of itching that arise from nothing more than itchy thoughts; dreams that can seem indistinguishable from reality; phantom sensations that amputees have in their missing limbs. And, the more we examine the actual nerve transmissions we receive from the world outside, the more inadequate they seem.

Our assumption had been that the sensory data we receive

from our eyes, ears, nose, fingers, and so on contain all the information that we need for perception, and that perception must work something like a radio. It's hard to conceive that a Boston Symphony Orchestra concert is in a radio wave. But it is. So you might think that it's the same with the signals we receive—that if you hooked up someone's nerves to a monitor you could watch what the person is experiencing as if it were a television show.

Yet, as scientists set about analyzing the signals, they found them to be radically impoverished. Suppose someone is viewing a tree in a clearing. Given simply the transmissions along the optic nerve from the light entering the eye, one would not be able to reconstruct the three-dimensionality, or the distance, or the detail of the bark—attributes that we perceive instantly.

Or consider what neuroscientists call "the binding problem." Tracking a dog as it runs behind a picket fence, all that your eyes receive is separated vertical images of the dog, with large slices missing. Yet somehow you perceive the mutt to be whole, an intact entity travelling through space. Put two dogs together behind the fence and you don't think they've morphed into one. Your mind now configures the slices as two independent creatures.

The images in our mind are extraordinarily rich. We can tell if something is liquid or solid, heavy or light, dead or alive. But the information we work from is poor—a distorted, two-dimensional transmission with entire spots missing. So the mind fills in most of the picture. You can get a sense of this from brain-anatomy studies. If visual sensations were primarily received rather than constructed by the brain, you'd expect that most of the fibres going to the brain's primary visual cortex would come from the retina. Instead, scientists have found that only twenty per cent do; eighty per cent come downward from regions of the brain governing functions like memory. Richard Gregory, a prominent British neuropsychologist, estimates that visual perception is more than ninety per cent memory and less than ten per cent sensory nerve signals.

The fallacy of reducing perception to reception is especially clear when it comes to phantom limbs. Doctors have often explained such sensations as a matter of inflamed or frayed nerve endings in the stump sending aberrant signals to the brain. But this explanation should long ago have been suspect. Efforts by surgeons to cut back on the nerve typically produce the same results that M. had when they cut the sensory nerve to her forehead: a brief period of relief followed by a return of the sensation.

Moreover, the feelings people experience in their phantom limbs are far too varied and rich to be explained by the random firings of a bruised nerve. People report not just pain but also sensations of sweatiness, heat, texture, and movement in a missing limb. There is no experience people have with real limbs that they do not experience with phantom limbs. They feel their phantom leg swinging, water trickling down a phantom arm, a phantom ring becoming too tight for a phantom digit. Children have used phantom fingers to count and solve arithmetic problems. V. S. Ramachandran, an eminent neuroscientist at the University of California, San Diego, has written up the case of a woman who was born with only stumps at her shoulders, and yet, as far back as she could remember, felt herself to have arms and hands; she even feels herself gesticulating when she speaks. And phantoms do not occur just in limbs. Around half of women who have undergone a mastectomy experience a phantom breast, with the nipple being the most vivid part. You've likely had an experience of phantom sensation yourself. When the dentist gives you a local anesthetic, and your lip goes numb, the nerves go dead. Yet you don't feel your lip disappear. Quite the opposite: it feels larger and plumper than normal, even though you can see in a mirror that the size hasn't changed.

The account of perception that's starting to emerge is what we might call the "brain's best guess" theory of perception: perception is the brain's best guess about what is happening in the outside world. The mind integrates scattered, weak, rudimentary signals from a variety of sensory channels, information from past experiences, and hard-wired processes, and produces a sensory experience full of brain-provided color, sound, texture, and meaning. We see a friendly yellow Labrador bounding behind a picket fence not because that is the transmission we receive but because this is the perception our weaver-brain assembles as its best hypothesis of what is out there from the slivers of information we get. Perception is inference.

The theory—and a theory is all it is right now—has begun to make sense of some bewildering phenomena. Among them is an experiment that Ramachandran performed with volunteers who had phantom pain in an amputated arm. They put their surviving arm through a hole in the side of a box with a mirror inside, so that, peering through the open top, they would see their arm and its mirror image, as if they had two arms. Ramachandran then asked them to move both their intact arm and, in their mind, their phantom arm—to pretend that they were conducting an orchestra, say. The patients had the sense that they had two arms again. Even though they knew it was an illusion, it provided immediate relief. People who for years had been unable to unclench their phantom fist suddenly felt their hand open; phantom arms in painfully contorted positions could relax. With daily use of the mirror box over weeks, patients sensed their phantom limbs actually shrink into their stumps and, in several instances, completely vanish. Researchers at Walter Reed Army Medical Center recently published the results of a randomized trial of mirror therapy for soldiers with phantom-limb pain, showing dramatic success.

A lot about this phenomenon remains murky, but here's what the new theory suggests is going on: when your arm is amputated, nerve transmissions are shut off, and the brain's best guess often seems to be that the arm is still there, but paralyzed, or clenched, or beginning to cramp up. Things can stay like this for years. The mirror box, however, provides the brain with new visual input—however illusory—suggesting motion in the absent arm. The brain has to incorporate the new information into its sensory map of what's happening. Therefore, it guesses again, and the pain goes away.

Not long ago, I met a man who made me wonder whether such phantom sensations are more common than we realize. H. was forty-eight, in good health, an officer at a Boston financial-services company living with his wife in a western suburb, when he made passing mention of an odd pain to his internist. For at least twenty years, he said, he'd had a mild tingling running along his left arm and down the left side of his body, and, if he tilted his neck forward at a particular angle, it became a pronounced, electrical jolt. The internist recognized this as Lhermitte's sign, a classic symptom that can indicate multiple sclerosis, Vitamin B12 deficiency, or spinal-cord compression from a tumor or a herniated disk. An MRI revealed a cavernous hemangioma, a pea-size mass of dilated blood vessels, pressing into the spinal cord in his neck. A week later, while the doctors were still contemplating what to do, it ruptured.

"I was raking leaves out in the yard and, all of a sudden, there was an explosion of pain and my left arm wasn't responding to my brain," H. said when I visited him at home. Once the swelling subsided, a neurosurgeon performed a tricky operation to remove the tumor from the spinal cord. The operation was successful, but afterward H. began experiencing a constellation of strange sensations. His left hand felt cartoonishly large—at least twice its actual size. He developed a constant burning pain along an inch-wide ribbon extending from the left side of his neck all the way down his arm. And an itch crept up and down along the same band, which no amount of scratching would relieve.

H. has not accepted that these sensations are here to stay—the prospect is too depressing—but they've persisted for eleven years now. Although the burning is often tolerable during the day, the slightest thing can trigger an excruciating flareup—a cool breeze across the skin, the brush of a shirtsleeve or a bedsheet. "Sometimes I feel that my skin has been flayed and my flesh is exposed, and any touch is just very painful," he told me. "Sometimes I feel that there's an ice pick or a wasp sting. Sometimes I feel that I've been splattered with hot cooking oil."

For all that, the itch has been harder to endure. H. has developed calluses from the incessant scratching. "I find I am choosing itch relief over the pain that I am provoking by satisfying the itch," he said.

He has tried all sorts of treatments—medications, acupuncture, herbal remedies, lidocaine injections, electrical-stimulation therapy. But nothing really worked, and the condition forced him to retire in 2001. He now avoids leaving the house. He gives himself projects. Last year, he built a three-foot stone wall around his yard, slowly placing the stones by hand. But he spends much of his day, after his wife has left for work, alone in the house with their three cats, his shirt off and the heat turned up, trying to prevent a flareup.

His neurologist introduced him to me, with his permission, as an example of someone with severe itching from a central rather than a peripheral cause. So one morning we sat in his living room trying to puzzle out what was going on. The sun streamed in through a big bay window. One of his cats, a scraggly brown tabby, curled up beside me on the couch. H. sat in an armchair in a baggy purple T-shirt he'd put on for my visit. He told me that he thought his problem was basically a "bad switch" in his neck where the tumor had been, a kind of loose wire sending false signals to his brain. But I told him about the increasing evidence that our sensory experiences are not sent to the brain but originate in it. When I got to the example of phantom-limb sensations, he perked up. The experiences of phantom-limb patients sounded familiar to him. When I mentioned that he might want to try the mirror-box treatment, he agreed. "I have a mirror upstairs," he said.

He brought a cheval glass down to the living room, and I had him stand with his chest against the side of it, so that his troublesome left arm was behind it and his normal right arm was in front. He tipped his head so that when he looked into the mirror the image of his right arm seemed to occupy the same position as his left arm. Then I had him wave his arms, his actual arms, as if he were conducting an orchestra.

The first thing he expressed was disappointment. "It isn't quite like looking at my left hand," he said. But then suddenly it was.

"Wow!" he said. "Now, this is odd."

After a moment or two, I noticed that he had stopped moving his left arm. Yet he reported that he still felt as if it were moving. What's more, the sensations in it had changed dramatically. For the first time in eleven years, he felt his left hand "snap" back to normal size. He felt the burning pain in his arm diminish. And the itch, too, was dulled.

"This is positively bizarre," he said.

He still felt the pain and the itch in his neck and shoulder, where the image in the mirror cut off. And, when he came away from the mirror, the aberrant sensations in his left arm returned. He began using the mirror a few times a day, for fifteen minutes or so at a stretch, and I checked in with him periodically.

"What's most dramatic is the change in the size of my hand," he says. After a couple of weeks, his hand returned to feeling normal in size all day long.

The mirror also provided the first effective treatment he has had for the flares of itch and pain that sporadically seize him. Where once he could do nothing but sit and wait for the torment to subside—it sometimes took an hour or more—he now just pulls out the mirror. "I've never had anything like this before," he said. "It's my magic mirror."

There have been other, isolated successes with mirror treatment. In Bath, England, several patients suffering from what is called complex regional pain syndrome—severe, disabling limb sensations of unknown cause—were reported to have experienced complete resolution after six weeks of mirror therapy. In California, mirror therapy helped stroke patients recover from a condition known as hemineglect, which produces something like the opposite of a phantom limb—these patients have a part of the body they no longer realize is theirs.

Such findings open up a fascinating prospect: perhaps many patients whom doctors treat as having a nerve injury or a disease have, instead, what might be called sensor syndromes. When your car's dashboard warning light keeps telling you that there is an engine failure, but the mechanics can't find anything wrong, the sensor itself may be the problem. This is no less true for human beings. Our sensations of pain, itch, nausea, and fatigue are normally protective. Unmoored from physical reality, however, they can become a nightmare. Doctors have persisted in treating these conditions as nerve or tissue problems—engine failures, as it were. We get under the hood and remove this, replace that, snip some wires. Yet still the sensor keeps going off.

So we get frustrated. "There's nothing wrong," we'll insist. And, the next thing you know, we're treating the driver instead of the problem. We prescribe tranquillizers, antidepressants, escalating doses of narcotics. And the drugs often do make it easier for people to ignore the sensors, even if they are wired right into the brain. The mirror treatment, by contrast, targets the deranged sensor system itself. It essentially takes a misfiring sensor—a warning system functioning under an illusion that something is terribly wrong out in the world it monitors—and feeds it an alternate set of signals that calm it down. The new signals may even reset the sensor.

This may help explain, for example, the success of the advice that back specialists now commonly give. Work through the pain, they tell many of their patients, and, surprisingly often, the pain goes away. It had been a mystifying phenomenon. But the picture now seems clearer. Most chronic back pain starts as an acute back pain—say, after a fall. Usually, the pain subsides as the injury heals. But in some cases the pain sensors continue to light up long after the tissue damage is gone. In such instances, working through the pain may offer the brain contradictory feedback—a signal that ordinary activity does not, in fact, cause physical harm. And so the sensor resets.

This understanding of sensation points to an entire new array of potential treatments—based not on drugs or surgery but, instead, on the careful manipulation of our perceptions. Researchers at the University of Manchester, in England, have gone a step beyond mirrors and fashioned an immersive virtual-reality system for treating patients with phantom-limb pain. Detectors transpose movement of real limbs into a virtual world where patients feel they are actually moving, stretching, even playing a ballgame. So far, five patients have tried the system, and they have all experienced a reduction in pain. Whether those results will last has yet to be established. But the approach raises the possibility of designing similar systems to help patients with other sensor syndromes. How, one wonders, would someone with chronic back pain fare in a virtual world? The Manchester study suggests that there may be many ways to fight our phantoms.

Appendix IV -- THE ENERGY CHALLENGE

The New York Times

Wind Energy Bumps Into Power Grid's Limits

By MATTHEW L. WALD

Published: August 26, 2008

When the builders of the Maple Ridge Wind farm spent $320 million to put nearly 200 wind turbines in upstate New York, the idea was to get paid for producing electricity. But at times, regional electric lines have been so congested that Maple Ridge has been forced to shut down even with a brisk wind blowing.

That is a symptom of a broad national problem. Expansive dreams about renewable energy, like Al Gore's hope of replacing all fossil fuels in a decade, are bumping up against the reality of a power grid that cannot handle the new demands.

The dirty secret of clean energy is that while generating it is getting easier, moving it to market is not.

The grid today, according to experts, is a system conceived 100 years ago to let utilities prop each other up, reducing blackouts and sharing power in small regions. It resembles a network of streets, avenues and country roads.

"We need an interstate transmission superhighway system," said Suedeen G. Kelly, a member of the Federal Energy Regulatory Commission.

While the United States today gets barely 1 percent of its electricity from wind turbines, many experts are starting to think that figure could hit 20 percent.

Achieving that would require moving large amounts of power over long distances, from the windy, lightly populated plains in the middle of the country to the coasts where many people live. Builders are also contemplating immense solar-power stations in the nation's deserts that would pose the same transmission problems.

The grid's limitations are putting a damper on such projects already. Gabriel Alonso, chief development officer of Horizon Wind Energy, the company that operates Maple Ridge, said that in parts of Wyoming, a turbine could make 50 percent more electricity than the identical model built in New York or Texas.

"The windiest sites have not been built, because there is no way to move that electricity from there to the load centers," he said.

The basic problem is that many transmission lines, and the connections between them, are simply too small for the amount of power companies would like to squeeze through them. The difficulty is most acute for long-distance transmission, but shows up at times even over distances of a few hundred miles.

Transmission lines carrying power away from the Maple Ridge farm, near Lowville, N.Y., have sometimes become so congested that the company's only choice is to shut down — or pay fees for the privilege of continuing to pump power into the lines.

Politicians in Washington have long known about the grid's limitations but have made scant headway in solving them. They are reluctant to trample the prerogatives of state governments, which have traditionally exercised authority over the grid and have little incentive to push improvements that would benefit neighboring states.

In Texas, T. Boone Pickens, the oilman building the world's largest wind farm, plans to tackle the grid problem by using a right of way he is developing for water pipelines for a 250-mile transmission line from the Panhandle to the Dallas market. He has testified in Congress that Texas policy is especially favorable for such a project and that other wind developers cannot be expected to match his efforts.

"If you want to do it on a national scale, where the transmission line distances will be much longer, and utility regulations are different, Congress must act," he said on Capitol Hill.

Enthusiasm for wind energy is running at fever pitch these days, with bold plans on the drawing boards, like Mayor Michael Bloomberg's notion of dotting New York City with turbines. Companies are even reviving ideas of storing wind-generated energy using compressed air or spinning flywheels.

Yet experts say that without a solution to the grid problem, effective use of wind power on a wide scale is likely to remain a dream.

The power grid is balkanized, with about 200,000 miles of power lines divided among 500 owners. Big transmission upgrades often involve multiple companies, many state governments and numerous permits. Every addition to the grid provokes fights with property owners.

The Revelation

These barriers mean that electrical generation is growing four times faster than transmission, according to federal figures.

In a 2005 energy law, Congress gave the Energy Department the authority to step in to approve transmission if states refused to act. The department designated two areas, one in the Middle Atlantic States and one in the Southwest, as national priorities where it might do so; 14 United States senators then signed a letter saying the department was being too aggressive.

Energy Department leaders say that, however understandable the local concerns, they are getting in the way. "Modernizing the electric infrastructure is an urgent national problem, and one we all share," said Kevin M. Kolevar, assistant secretary for electricity delivery and energy reliability, in a speech last year.

Unlike answers to many of the nation's energy problems, improvements to the grid would require no new technology. An Energy Department plan to source 20 percent of the nation's electricity from wind calls for a high-voltage backbone spanning the country that would be similar to 2,100 miles of lines already operated by a company called American Electric Power.

The cost would be high, $60 billion or more, but in theory could be spread across many years and tens of millions of electrical customers. However, in most states, rules used by public service commissions to evaluate transmission investments discourage multistate projects of this sort. In some states with low electric rates, elected officials fear that new lines will simply export their cheap power and drive rates up.

Without a clear way of recovering the costs and earning a profit, and with little leadership on the issue from the federal government, no company or organization has offered to fight the political battles necessary to get such a transmission backbone built.

Texas and California have recently made some progress in building transmission lines for wind power, but nationally, the problem seems likely to get worse. Today, New York State has about 1,500 megawatts of wind capacity. A megawatt is an instantaneous measure of power. A large Wal-Mart draws about one megawatt. The state is planning for an additional 8,000 megawatts of capacity.

But those turbines will need to go in remote, windy areas that are far off the beaten path, electrically speaking, and it is not clear enough transmission capacity will be developed. Save for two underwater connections to Long Island, New York State has not built a major new power line in 20 years.

A handful of states like California that have set aggressive goals for renewable energy are being forced to deal with the issue, since the goals cannot be met without additional power lines.

But Bill Richardson, the governor of New Mexico and a former energy secretary under President Bill Clinton, contends that these piecemeal efforts are not enough to tap the nation's potential for renewable energy.

Wind advocates say that just two of the windiest states, North Dakota and South Dakota, could in principle generate half the nation's electricity from turbines. But the way the national grid is configured, half the country would have to move to the Dakotas in order to use the power.

"We still have a third-world grid," Mr. Richardson said, repeating a comment he has made several times. "With the federal government not investing, not setting good regulatory mechanisms, and basically taking a back seat on everything except drilling and fossil fuels, the grid has not been modernized, especially for wind energy."

Appendix V -- Lexicon

Advanced Nurturing: The theory presented in this book describing the stage of evolutionary development where animals care intensively for their young. It is the precursor to the C-brain adaptation in humans.

BHU: Also see CHU. The minimum a person must produce and consume in order to survive for one year (including the person themselves). When an individual produces more than 3 times the amount of a BHU, he/she becomes a CHU, or Competitive Human Unit. CHUs have a significantly more positive effect upon their economy (gas cloud) than do BHUs.

Black Hole: also the Singularity, a collapsed supernova, a nearly perfectly compressed sphere, a separate and infinite universe where time and space do not apply, God.

Brain Dependency-Deficiency: each human is just ½ of a complete man/woman or "mated" brain unit. If two variables are more productive because they job share—like a wrench and hammer—then it is a mathematical certainty that one can do one job well and *not be* as suited to the job the other does well. Separate the tools, and this demands that each tool will now have things they don't accomplish at maximum efficiency.

Brutal/subsistence experience: the vast majority of time during which humanity was just able to survive. This time period corresponds to most of time here on Earth actually. See Indulgent Experience also.

C-Brain: the Conscious mind is contained in the smaller, weaker, "blank slate," tool-making, calculating, and socializing brain, the C-brain. The C-brain is located physically on the top of the brain, mostly in the cerebral cortex, and is only 20 to 30 percent of the actual brain mass. The C-brain is home to logic, calculation, reason, charity and intellectual work.

CHU: An individual producing three times, or more, of a BHU, becomes a CHU, or Competitive Human Unit. Also see BHU. The CHU portion of the world economy is vastly bigger than the BHU portion due to the productivity multiplier afforded by education, physical opportunity and access to investment capital. The sum of the CHU contributions to the economic equation is what we call GDP, the monetized portion of the economy.

Cortical Brain: the top 20% of the human brain, responsible for most thought we think of as "Conscious" thought. See also C-brain.

Combined Laws of Thermodynamics, (CLOT): these elegant principles laid down by Max Planck describe the limitations and characteristics of energy in a gas cloud. The theory presented in this book states that these principles underlie all the activity in our monetized economy and actually dictate the behavior of better-known principles like supply and demand.

D-brain: the larger, older, evolution-shaped/influenced part of the brain that controls the base functions and emotions as well as thousands of complex programs inherited from generations of evolution which also control and/or influence behavior. Humans are in denial about the existence of the D-brain because the human consciousness mostly resides in the other layer, your C-brain, or conscious brain. The D-brain is 70 to 80 percent of the brain by volume and is located, not on the left or the right side, but physically *beneath* the cerebral cortex, the upper-most layer of the brain, the one containing most of the C-brain. D-brain is very focused on maintaining a constant vigil against death, opposed/tempered only by a strong impulse to have sex and procreate. This *base layer* of our brain makes most of the important decisions in a human's life. It is home to denial, love, hate, rage, racism, greed, satisfaction, and most importantly, evolutionary behavior patterning—the real secret to understanding human behavior.

Denial: the ability to lie to oneself; this requires a second consciousness as described in this book. See also Mass Denial.

Directed Mutation: the theory, presented in this book, that prior to the evolution of higher order creatures, the first of several types of feedback mechanisms that service the cell at the molecular level randomly developed a way to influence the changes (mutations) in a certain *direction*. This *extra* mechanism (random mutations also concurrently occur) is driven by environmental factors. Hence the mutations occurring are often directed by changes in the environment—affecting the organism's ability to exploit the new environment in a positive manner more quickly, more precisely, and through more than one adaptation at a time.

Devil's Advocate: before its abolishment in this century, an official post in Rome. The Bishop or Cardinal assigned to research a proposed Saint and acting as *The Devil's Advocate* would gather the necessary evidence and testimony needed to discredit the miracles attributed to the proposed Saint.

Eve: humanity's common female ancestor as the anthropologists like to call her, she lived just 160,000 years ago and by DNA evidence is the oldest common ancestor to all living humans.

Fundamental: Fundamentals are large evolutionary strategies that push adaptations out in front of them. Fundamentals are the most basic force of biologic evolution. They cross many species boundaries and represent the aggregate of the biologic evolutionary direction at any time in history. Currently, the big and strong Fundamental is out. It was wiped out by a meteor strike 65 million years ago. To picture a Fundamental, imagine the water gently flowing past the blunt bow on a slow moving boat. One such Fundamental used by the dinosaurs was a combination of strength and size. In the carnivorous dinosaurs these were combined with speed, ferocity, biting power and killer instinct. The Fundamental involved in man's ascension to today's advanced consciousness is advanced nurturing and/or love.

GDP or Gross Domestic Product: The sum of everything humans make or do; our entire work product, whether it's cut someone's hair, build a car, or grow vegetables. This can also be expressed as wealth, since it is the change in GDP from year to year that determines whether the economy is "growing" or not.

Human Inherited Memory: according to the theory presented in this book, is knowledge passed down generationally in each human brain from millions of years of evolutionary trial and feedback. This data according to our theory, is readily accessible to the brain, but not by the conscious mind directly. It is not just knowledge; it's morals and emotions as well.

Indulgent experience/environment: an economy that provides so much productivity and wealth for it's members, that the time spent on survival needs--to be assured of food, fuel and shelter -- is so low that the C-brain has excess time for planning, advanced nurturing and ideas that would normally be dangerous to the organism's survival.

Killer: a term used throughout this book to describe man, mankind and human activity in general. Killers frequently kill to achieve their economic and survival goals, even though in humans we mask the real reasons for this behavior with denial and lying.

Manipulation: the theory presented in this book removes the negative emotional connotation that normally exists with this word. C-brains use Manipulation in every vocal utterance, every physical effort, clearly attempting to instigate in the humans around them reciprocation, material payment, and/or emotional payment (e.g. love, praise, flattery, respect, attention etc.,) for their co-participation in the social exchange. It is the vast number of psychological, physical, emotional, and mental actions, thoughts, words, and deeds humans develop in their conscious mind to manipulate the humans in the world around them into doing what they want the other person to do. Materially speaking, this is highly-productive behavior.

Marshall Plan: the re-construction plan for Europe and Japan carried out largely by the U.S. government after WWII.

Mass Denial: Agreed upon lies human tell in groups and believe.

Non-Power Sharing: Populations: otherwise known as Communist States, Dictatorships, Dynasties, Monarchies, etc. The definition in this book is broader, and includes societal groups of people whose C-brains have not become developed enough to desire the responsibility for selecting their laws and leaders. This lack of democratic interest can be due to lack of education or lack of sufficient food and shelter. Even normally Democratic populations *desire* less power sharing with their government during troubled environmental times.

Non-Singularity Universe, (NSU): the universe created by the Singularity. This universe is where we humans reside. It is a two variable universe or free will possible universe.

One (1): also the Singularity. Also: a black hole, the first number, God, and an entity that can only be divided by itself, thereby demanding consciousness. Also see Black Hole and Singularity.

Peace Machine: the human brain's unique architecture and the resultant unique routing of knowledge and memory flow so as to exert control over human behavior and create a "Social Brain" (in essence, a *Peace Machine*). The Social brain allows predator/killer humans to live together in groups with less bloodshed. Without it, the gains from living together wouldn't be worth it evolutionarily speaking.

Power-sharing populations: otherwise known as Democracies. The definition is broader in this book, so as to describe the groups of people whose C-brains have become developed enough to want input into the decision making process for their laws and leaders.

Racism: the ingrained fear of different looking humans. It is biologically programmed in the D-brain.

Revelation: a multi-dimensional memory without words or letters, containing ideas and images fused together. Its components are composed of images that interact with each other in ways that physically demonstrate mathematical and logical relationships. A three-dimensional space that a mind can wander into, full of understanding displayed as ideas, equations, and images--but no text or sound.

Singularity: a universe that is infinite and not subject to space and time. Also describes: God, a black hole, the *original* universe, the number One (1), and the creator (by division) of the Non-Singularity Universe we live in. See also Black Hole, God, Non-Singularity and One (1).

Super Darwinism: the theory, presented in this book, that man has two brains. One evolved to enable him to kill other creatures and survive, one evolved later, as a tool of the killer brain, to get along with other humans. The impetus for this later adaptation was love. The evolutionary benefit was better survival in groups.

Swinging Door Principle: an element that must be incorporated into a world constitution that differs from our U.S. Constitution, as well as virtually all other, world models. This principle would call for each member state to be responsible for its own country's defense, and provide for each nation-state to be allowed relatively easy transition in and out of the Federation.

Tabula Rasa: (Blank Slate) is the philosophy that our brains are born blank and our mind is written upon by experience, which in turn influences our decisions and morals. John Locke, who, along with Rousseau, heavily influenced the thinking of the United States' Founding Fathers, first proposed it. This book explains how this describes only our C-brain.

Unified Brain Command: the theory, presented in this book, that the C & D brains are not obvious to humans because a unified brain command is necessary for survival. Thus, we have evolved powerful denial and lying mechanisms to prevent a split decision and the resulting confusion, panic, and death of the organism.

Unified Field Theory: a theory presented in this book that explains the interconnection of God, physics, and the world we live in right down to the atom.

Predictions and Recommendations

1. Universal use of robots will be the next big transformation. Companies such as Honda are pioneering this field. This prediction is based upon the productivity principle. Robots are incredibly productive; they can work many more hours than a human; they can do physically harder jobs; they do not require a medical plan; they don't want holidays. And, while people are born liars, robots don't lie. Not only do robots raise productivity levels dramatically, they also do the job with a minimum of energy - very important.

2. New materials are critical to our future, especially super-strong lightweight materials. The innovations to first occur will be lightweight metal alloys that are lighter and stronger than the current standard metals, yet are cost-effective. Buy stock in companies that bring new, improved materials manufacturing to market--it will help them and you.

3. Computers will remain a growth field. Computers will get smaller, smarter, and more capable. Learn computer skills, and use computers to control your environment. Buy stock in companies that produce computers that offer significant productivity advantages.

4. Oil is "out" but will not die without a struggle. Though the oil supply continually contracts for the next thirty years, demand will not fall fast enough. This means it will remain a "big money game" with large incentives to stifle all alternatives. Energy alternatives are the future; make sure they are working for you, or you will not be working. Simply forcing yourself to include alternative energy supplies in your business plan could be the edge that saves your business.

5. Globalization is not just an expression. The drugs a doctor injects into your family may come from China, India, Japan, or Europe. The tainted products made in China that were headline news recently are one example of the difficulties we face with globalization. People need safe products, produced efficiently. Wars, especially civil wars, will inevitably happen as D-brain-heavy populations achieve the wealth and education level necessary to support C-brain power-sharing schemes like democracy and/or capitalism.

6. More education will mean more wealth for you. More education for everyone will mean more wealth for everyone. To motivate your

actions by love, you need to be concerned with both of these issues. While you may be committed to the first more so than the latter, your commitment to both must be absolute. Commitment to improving only yourself will eventually fail you.

7. The more you love your children, the more they will be capable of love – for you -- and others. Whatever love we input, gets returned times ten.

8. Manners and ethics will serve you—if you use them.

9. God is Love; Love is God. The Revelation placed this at its core.

10. ***God cares not so much about what you do as He does about your motivation.*** *If your actions are motivated by love, you will be carrying out God's plan.*

Conjecture and *Angels*

There are thoughts, especially at the conclusion of a book like this, that occur during writing and have no place in the narrative. I've jotted a few down, at this place I call conjecture. To really love someone and motivate your thoughts for that person by love, you must be willing to sacrifice for that person, and to work productively to give them the fruit of that labor. In times of self-defense, you must be willing to fight for their defense *and die in their defense*. Motivating your actions by love doesn't mean you live forever in the NSU, just in the SU!

If you dream good thoughts of your family, visualizing them happy and strong, you will make better decisions and their happiness will be realized. You cannot fix the world—that is love's job. Choose love; use love as a tool to build your family. The rest will always work itself out, though not necessarily in the way you planned or hoped. This may be difficult to understand.

In other words, your self-created expectations will always fall short of God's complete plan for you. He has perfect knowledge of your circumstances (and beyond) and you cannot.

If you simply allow Him to guide your path as you act with love instead of trying to fully direct your own life, you will find that His results will always exceed your plan. Frankly, we can easily see this problem in others if we objectively observe them. When others begin grasping after failing to meet their expectations, we instinctively know their actions are wrong and will fail. When we act this way, our denial mechanism blocks our self-perspective.

So, work hard, regardless of your occupation. But don't grasp. Let love motivate your decisions, and far greater things will be accomplished by you than you would ever imagine possible. God doesn't just have a better imagination than you; He has one connected to all the other moving parts of the universe. Since productive work is part of love, any worker can channel love (God) through his or her work. While we notice this channeling more in artists than in our auto mechanic, the principle is identical in every worker.

Once you learn to see Love/God peeking out through people, you will recognize the angels among us. These are people who motivate their actions with love a majority of the time toward those around them. They quietly push friends and family forward, mostly women to be honest, always helping others when someone is sick or in need. This process, by the way, will teach your brain to begin seeing love-motivated actions in others.

The human Angels can be differentiated from the manipulators around us like grain from the chaff. You can learn to recognize them. An angel will not want you powerless. Nor will his or her generosity of spirit, money, advice, and emotion be designed to make you dependent or compromised after the association. *This is a result of the Angel's motivations.*

To receive love is healthy for you. It should not be motivated by a desire to also weaken you in *any* way. Many co-dependent people use "Love-motivated-appearing behavior" as a manipulator's cloak for self-interest and self-dependency. In associating with these false angels, you can feel the unhealthiness most strongly at the moment of separation. These manipulators hang on to make sure the manipulation achieved its desired result. Conversely, you will never feel an Angel separate. The Angel does it without extracting a cost; the sole aim is that you use the benefits of love, become strengthened by it, and grow strong enough to love others in that way. Look for Angels. When you find them, cherish them.

For some reason, Angels simply don't come along as often in our adult lives as when we were children. They most often direct their love at children. One day, when you look around and remember the pureness of their thoughts, you will miss them. I no longer cry in pain about the past; but I still cry a lot. It's more in wonder. Now I know where past moments of loving behaviors exist… beyond my memories. They are in the Singularity because they were motivated by love.

Best regards and blessings, the SU's humble servant, Karl Spain.

Made in the USA
Lexington, KY
26 October 2010